Sorin Vlaicu

Spinning Ring Electron

2017

FOREWORD

Soon after J. J. Thomson proved experimentally in 1897 that the electron is not only an undefined quantum of electricity, but also an infinitesimal corpuscle with determinate mass of about $9.1 \cdot 10^{-31}$ kg, a surprising property of this first elementary particle became increasingly more evident: in addition to the electric field with spherical symmetry generated by the point-like elementary electric charge e attached to each electron, and even contradictory to this field, any electron generates in the surrounding space its own dipole magnetic field with axial symmetry, similar as shape and equation to those generated at macroscopic level by all stationary electric currents of circular form. That is why in the second decade of the past century more and more physicists were becoming maintainers of a spinning ring electron with linear peripheral velocity equal or very close to the already known speed of light $c = 3 \cdot 10^8$ m/s and size of about a hundred times smaller than hydrogen atom. And these estimations were consolidated in the first years of the third decade, when Stern and Gerlach obtained a relatively accurate magnetic moment $M_e \approx 9 \cdot 10^{-24}$ A·m² of the electron by using a new measuring method. Also, many other researchers confirmed a good compatibility between the classical structural model of spinning ring electron and their experimental data in different domains.

For all that, since the last few years of the same decade this classical model of spinning ring electron was gradually abandoned as incompatible with other theories recently elaborated by great personalities of much account, as Einstein and Bohr. For instance, if the spinning ring electron with peripheral linear velocity c has rest energy $E_0 = m_0 c^2/2$, where m_0 is the rest mass of the particle, conforming to Einstein's special relativity the rest energy of the electron is $E_0 = m_0 c^2$. Likewise, if the spinning ring electron with electric charge e is a circular electric current of radius r_e, magnetic moment $M_e = ecr_e/2$, angular momentum (or spin) $L_e = m_0 c r_e$ and gyromagnetic ratio $g_e = M_e/L_e = e/2m_0$, as in the meanwhile its magnetic moment M_e was found to be equal to one Bohr magneton $M_B = eh/4\pi m_0$, where h is the Planck's constant, its own angular momentum had to be $L_e =$

$h/2\pi = \hbar$. But in 1925, in a time when the Bohr's atomic theory proved to be unable to explain the recently discovered fine structure of hydrogen spectral lines, Goudsmit and Uhlenbeck found that the fine splitting of hydrogen spectral terms could be somewhat justified if the electron would have an angular momentum twice smaller $L_e = h/4\pi$, therefore a gyromagnetic ratio correspondingly larger $g_e = e/m_0$, and their speculation was enough for accepting the *semi-integer* spin $L_e = \hbar/2$ of the electron, even if their theory includes as an essential point an *orbital magnetic moment* of the electron in hydrogen atom, a fabrication in total disagreement with the basic laws of electromagnetism.

Indeed, according to these laws, magnetic moment is a physical quantity proper exclusively to the dipole magnetic fields, and even if the orbital motion of an atomic electron can be seen formally as an electric current, a single electron in circular motion is definitely unable to generate a dipole magnetic field, whose existence is strictly conditioned by a uniform spreading of the whole electricity quantity of the stationary electric current on its plane circumference. Just in this case all Biot-Savart magnetic fields simultaneously generated any minute from each point of this circumference result by vector summing in a dipole magnetic field defined by a magnetic moment $M = IA$, where I is the constant intensity of the electrical current and A is the plane area delimited by its circumference.

More, conforming to Goudsmit and Uhlenbeck this fictitious dipole magnetic field generated by the orbital motion of an atomic electron is able to alter even the energy of its elementary source!, what is another monumental aberration, because any electron in motion generates a magnetic field whose force lines never intersect it. In fine, for unknown reasons such an effect does not take place in the s states of the atoms, although the orbital motion of atomic electrons has evidently to exist even in these atomic states!

But the spinning ring electron has been undesirable for special relativity not only for its rest energy $E_0 = m_0 c^2 / 2$ different from relativistic rest energy $E_0 = m_0 c^2$, because any kind of electron with internal structure is absolutely incompatible with special relativity in many other respects, so that the only structural model of electron compatible with special relativity is the so-called *pointlike* electron, seen as a non-dimensional entity, an ultimate brick of the Universe.

On the other hand, though, how could such a pointlike electron with no internal structure have angular and magnetic own moments, which can be defined only for electrically charged bodies in rotation? But as nothing was impossible in the *crazy* years 1920's, Dirac found out in 1928 his famous relativistic equation of the pointlike electron, conforming to which the pointlike electron should really have an *intrinsic* angular moment $L_e = h/4\pi$ and *intrinsic* magnetic moment $M_e = eh/4\pi m_0$, and so the two *intrinsic* moments of the pointlike electron became *quantum relativistic effects* only similar to those defined by the laws of physics, but in fact with no connection with a rotation motion of a composite electron.

Although initially the Dirac's theory was very reticently received owing to its extravagant ideas, full of contradictions and hard digestible even in those *revolutionary* years, finally the Einstein's adherents had no choice and accepted it as a theoretical ground for the relativistic pointlike electron. Moreover, this understanding of the two moments as *quantum relativistic effects* is used even today, although in the meantime essential predictions of Dirac's theory were experimentally invalidated as respects the anomalous magnetic moment of the electron and the Lamb shift.

Not much later, the rotating ring electron with classical rest energy $E_0 = m_0 c^2/2$ got an ultimate blow in 1934, when Klemperer found experimentally the rest energy about 0.5 MeV of the electron, which corresponds to relativistic formula $E_0 = m_0 c^2$ of this energy, so that from that moment the pointlike electron has remained the only passable version, in such circumstances any doubt in this respect becoming vain.

However, until now someone ought to notice that all quantum relativistic formulas proper to the rest energy, gyromagnetic ratio and angular moment of the relativistic pointlike electron, in order

$$E_0 = m_0 c^2 \qquad g_e = e/m_0 \qquad L_e = h/4\pi,$$

differ from their homologous formulas of the classical spinning ring electron,

$$E_0 = m_0 c^2/2 \qquad g_e = e/2m_0 \qquad L_e = h/2\pi,$$

by the same 2-factor, and, what is much more important, if the real value of the Planck's constant would be exactly twice smaller than that always considered $h = 6.626 \cdot 10^{-34}$ J·s, the previous three equations of the pointlike electron would be invalidated, and those of the spinning ring electron become suddenly correct! Or, even just a hypothetical possibility for such a dramatic overturn in questions fundamental for the whole physics of microcosm (or physics of elementary particles) would have required an exhaustive review of the whole history of the action constant, especially when some experiments as magnetic resonance or Josephson transitions needed theoretical innovations out of the basic laws of physics, and even out of an elementary logic, exactly for introducing a saving 2-factor in their energy equations, without which their experimental data would have lead to a halved Planck's constant $h = 3.313 \cdot 10^{-34}$ J·s !

Well, such a critical review of all the really relevant experiments performed along the time for measuring the Planck's constant is at last done in the first chapter of this book, and all reexamined experiments confirm a clear overvaluation of the Planck's constant always used until now in theoretical and experimental physics of microuniverse, exactly with that 2-factor able to decide between the classical ring electron and the pointlike electron imposed by quantum relativistic physics. Indeed, both experimental methods used in the 1910's for measuring the Planck's constant prove to have wrong energy equations, whose unquestionable omissions have doubled the real value of this constant, and this conclusion is decisively reinforced by the inadmissible manner in which the experimental data resulted later from the other two essential experiments previously mentioned were deliberately distorted for avoiding at all costs the necessary halving of a constant whose 1/2 reappraisal would have meant a catastrophic collapse of all recent and much glorified quantum relativistic theories, together with an obligatory return to classical physics of microcosm.

Evidently this late halving of the Planck's constant means before anything else the reinstatement of the classical model of spinning ring electron unjustly removed long ago, but here this restoration goes much further by its thoroughgoing development in connection with the older idea about the existence of a universal matter out of which all the known elementary particles are shaped in various forms. And finally this more advanced structural model of spinning ring electron proves not only to be in agreement with numerous experimental data appeared after its interested abandonment, but also to be very productive for deducing structural models for all the other main elementary particles, able in turn to explain all essential properties of these particles, some of them unsolvable mysteries for quantum relativistic physics.

CONTENTS

ORIGINAL MISTAKE OF QUANTUM PHYSICS

How electromagnetic radiation was quantified unwillingly 7
Revising first experimental measurements of action constant 8
Electronic magnetic resonance 15
Josephon transitions 19
No other self-consistent measurements of action constant 21
Return to spinning ring electron 23

NOTIONAL ATOMIC HYDROGEN

Continuous end of spectral series assigned to atomic hydrogen 27
Fine and hyperfine splitting in hydrogen line spectrum 28
Strange case of orthohydrogen magnetic moment 36
Microwaves $\lambda = 0.211$ m from hydrogen maser and cosmic space 41
Too many grave discrepancies ignored for too long 47

ELEMENTARY EVIDENCE AGAINST SPECIAL RELATIVITY

Nuclear and atomic mass defects 49
Electron with variable rest mass and energy 55
Free electrons accelerated in electromagnetic field 57
No indisputable proof for invariant velocity of light 59
Experimental test in reality experimental denial 68
Postulate for an imaginary space 70

SPINNING RING ELECTRON

Brief history 73
Ring electrons as infinitesimal circular electric currents 79
Ring electrons in extremely strong magnetic fields 85
Magnetic nucleons 90
Ring electrons accelerated in electromagnetic field 96
Wavicle electron, reality or fantasy much too easy accepted? 98

PHOTON EMISSION OF RING ELECTRON

Beam photon 103
Corpuscular beam photon instead wavicle photon 109
Photon emission of differently polarized ring electrons 114
Polarized beam photons 120
Spiral neutrinos radiated through multiphoton emission 124
From Newton's particles of light to preons 133

FOUR FUNDAMENTAL INTERACTIONS

Universal weak interaction 135
Strong interaction between coaxial ring quarks 138
Gravitational interaction 144

RING ELECTRON WITHOUT ELECTRIC CHARGE

Magnetic but neutral ring electron 153
Electric or magnetic fields? 155
Forces acting on electrons in magnetic fields 160
Attractive interactions between electrons 163
Atoms made of nuclei and electrons without electric charges 165

ORIGINAL MISTAKE OF QUANTUM PHYSICS

How electromagnetic radiation was quantified unwillingly

Even if first quantification in physics was undoubtedly that one of electricity, whose granular structure was stated by Franklin in 1749, and already in 1874 Stoney succeeded even in calculating an elementary electric charge of about 10^{-20} C magnitude order by using the Faraday's laws of electrolysis, the real birth certificate of quantum physics has been almost unanimously placed in 1900, when Planck was compelled to postulate that the energy radiated by a blackbody is not a continuous variable, as it has been assumed until then, but a sum of a big number of discrete elements of energy $\epsilon = h\nu$, where ν is their frequency and h is an action constant. And although at the beginning Planck saw his postulate $E = n\epsilon = nh\nu$ (where n is an integer number) just as "a purely formal assumption", and he even proposed to eliminate this undesirable quantification of electromagnetic energy as quickly as possible, this was the first step on a long way whose end has been the present quantum and relativistic physics.

In fact Planck resorted to this quantification of electromagnetic energy — seen by him as "an act of despair" by which he was obliged "to sacrifice any of my previous convictions about physics" — because only by adopting this new approach to the concept of energy his older law of blackbody radiation could be brought into accord with experimental data in the whole range of electromagnetic radiation emitted by hot bodies. And such a resounding success could not be refused in that age, when the other two known laws of blackbody radiation had distinct domains of applicability: the Rayleigh-Jeans law valid only for the high frequencies emitted by bodies heated at high temperatures, and the Wien law only for the low frequencies emitted by colder bodies.

More, already in the next years the Planck's postulate $E = nh\nu$ proved to be a very effective implement in the whole physics of microcosm, and not at all a merely mathematical formalism, as Planck still was claiming. Its first important new application appeared in 1907, when Einstein postulated as based on the Lenard's discovery of photoelectric effect that the light itself consists of discrete quanta, which later were recognized as distinct elementary particles called photons and acknowledged as the field particles of all electromagnetic interactions. But such energy quanta with mass and momentum, like any macroscopic body!, were already too much for Planck, who rejected in spite of evidence this new corpuscular theory of light, and implicitly of the whole electromagnetic radiation, only because this corpuscular variant could not be integrate in Maxwell's electrodynamics, much too deeply rooted in his mind to be doubted one way or another.

Other important applications of the concept of electromagnetic quanta were particularly atomic theory proposed by Bohr in 1913, and later first relativistic kinematics thought out by Compton in 1923. Also, in its reduced form $\hbar = h/2\pi$ the Planck's constant proved to be a natural unit for measuring the proper kinetic moments of all elementary particles, which has determined its presence in many fundamental equations of particle physics.

Evidently, this gradual extension of the Planck's constant in the whole physics of microcosm has made essential its numerical value. Planck himself gave the first numerical value of his action constant in 1901, when he calculated $h = 6.55 \cdot 10^{-34}$ J·s based on the laws of blackbody radiation and some thermodynamic data available at that time. And because this first size needed some independent acknowledgments, two experiments were carried out to this end in the second decade of the past century. Both of them confirmed the Planck's value, but now their attentive reexamination changes this soothing conclusion.

Revising first experimental measurements of action constant

In 1916 Millikan[1] used a beam of monochromatic light with known frequency ν in order to obtain photoelectrons from a metal lattice containing a big number of quasi-free electrons whose binding energy in atoms and extraction work from the lattice were negligible with respect to the energy $h\nu$ of the incident photons absorbed by them. In this case the whole energy $h\nu$ of an absorbed photon is practically converted into the kinetic energy $E_k = m_0 v^2/2$ of the emergent photoelectron, $h\nu = m_0 v^2/2$, where m_0 is the

---

[1] R. A. Millikan, *A Direct Photoelectric Determination of Planck's "h"*, Phys. Rev. **7**, (1916) 355.

mass and v is the speed of the resulted photoelectron (the classical kinetic energy $E_k = m_0 v^2/2$ and the rest mass $m_0 = 9.109 \cdot 10^{-31}$ kg of the electron can be used with enough accuracy when the speed v is small in comparison with that of light, $v \ll c$). Finally, the monoenergetic photoelectrons with velocity $v = \sqrt{2h\nu/m_0}$ were brought back at their initial rest state by stopping them in a delaying electromagnetic field with known potential V_{br}, and the action constant was calculated by Millikan from equality

$$h\nu = eV_{br},$$

from which it resulted a numerical value $h = 6.547 \cdot 10^{-34}$ J·s almost identical to that calculated by Planck in 1901 for his action constant.

In after years several researchers in United States and Germany used a more accurate method, first time carried out by Blake and Duane[2]. At first the free electrons at relative rest on the cathode of an electronic gun were accelerated under a known voltage V to a relatively small velocity $v \ll c$ determined by equality $eV = m_0 v^2/2$ (this total conversion of the energy eV transferred by an electromagnetic field to the electrons into their kinetic energy $E_k = m_0 v^2/2$ was established by J. J. Thomson since 1897). Subsequently these monoenergetic electrons were braked within the anticathode by successive inelastic collisions with atomic electrons more or less free in the metal lattice, finally up to their state of relative rest. Because during each of its successive braking stages in the metal lattice a free incident electron emits a photon whose energy is equal to its energy decrease in that stage, until its return to the final state of relative rest such an electron radiates a variable number of photons with different frequencies, so that on aggregate the bremsstrahlung evolved in anticathode by all the incident electrons has a continuous spectrum. And because the upper limit frequency ν_{max} of this spectrum corresponds to a braking of the accelerated electrons in only one stage, directly from their final velocity v to their rest state, the Planck's constant was determined from the energy balance

$$eV = h\nu_{max},$$

and the resulted value was $h = 6.624 \cdot 10^{-34}$ J·s.

Since the two different methods thought for measuring the Planck's constant given results similar not only between them, but also similar to the

---

[2] F. C. Blake and W. Duane, *The Value of "h" as Determined by Means of X-Rays*, Phys. Rev. **10**, (1917) 624.

value calculated by Planck himself, and this numerical value of the action constant was meanwhile used with a resounding success by Bohr in his atomic theory, everybody was satisfied. But yet, still since then two inadmissible errors in the energy equations used in these experiments should have had to be noticed and corrected.

First of them, very easy noticeable, but amazingly overlooked by all even until today, it was the absence of the bremsstrahlung in the energy equation used by Millikan. Indeed, in both methods the electrons were first time accelerated by the agency of the photons — either the ones in a light beam incident in a metal lattice or those carrying the electromagnetic interaction in an electromagnetic field generated by oscillating currents of high frequency — from a relative rest state to a small velocity $v \ll c$ determined by the energy equations $h\nu = m_0v^2/2$ and, respectively, $eV = m_0v^2/2$, but after this first stage of positive acceleration through photon absorption, in both experiments the moving electrons were brought back at a relative rest state — either in a delaying electromagnetic field whose photons move in directions opposite to those of the electrons, as in Millikan experiment, or by their braking within an atomic lattice, as in the other method — but such a stopping process is ever accompanied by a photon emission called bremsstrahlung. For all that, until now nobody has wondered why this braking radiation simply did not exist in the Millikan's balance of energy, although it is the key element in the energy balance of the other method used for measuring the Planck's constant about the same time! And it must be said clearly, there was no excuse for this omission even at that time, because already in 1901 Kaufmann ascertained experimentally that the mass of the β-electrons with velocities $0.7 \ldots 0.9\ c$ increases as their velocity is higher, and both Lorentz in 1904 and a little later Einstein in 1905 proposed the same equation for the dependence of the electron mass upon its velocity,

$$m = m_0/\sqrt{1 - v^2/c^2},$$

an equation in good agreement with Kaufmann's experimental results. More, after eight years Bücherer confirmed with a higher accuracy this mass equation for β-electrons too, but with smaller speeds $0.3 \ldots 0.53\ c$, and in 1915 Guye and Lavanchy confirmed it again, even more accurately, for the electrons accelerated in electromagnetic field to velocities $0.26 \ldots 0.48\ c$. Or, if all the electrons electromagnetically accelerated to velocity v have an increased mass $m = m_0/\sqrt{1 - v^2/c^2} > m_0$, their return to the rest mass m_0

involves inevitably a photon emission of them, no matter how they are braked, because the bremsstrahlung is the only way by which the electrons can get rid of their motion mass $\Delta m = m - m_0 = m_0 \left[1/\sqrt{1 - v^2/c^2} - 1\right]$, or $\Delta m = m_0 v^2/2c^2$ for small velocities $v \ll c$. And even if in both experiments this mass decreasing Δm is very small as against the rest mass m_0 of the electrons, $\Delta m \ll m_0$, the bremsstrahlung E_{br} radiated during the braking processes has obviously a magnitude order which excludes its absence in the energy balance of the Millikan's experiment.

On the other hand, although the bremsstrahlung energy E_{br} has unquestionably to be included in the energy balance of the the Millikan's experiment, this obligatory correction raises another problem: a new term in the energy balance would change significantly the numerical value resulted in this experiment for the action constant, so that the two methods used in the 1910's for measuring this constant would lead to different numerical values, which is also unacceptable. Therefore, the absence of the bremsstrahlung energy E_{br} in the Millikan's energy equation is certainly not the only mistake that must be corrected, because the two experiments have to give the same numerical value of the Planck's constant if both their energy equations are correct.

And indeed, another fundamental error can easy be noticed in both methods used in the 1910's for measuring experimentally the action constant: the photon energy E_γ radiated outside by the electrons during their electromagnetic acceleration from the rest state to a certain velocity v was entirely ignored. And again there was no excuse for this second grievous mistake even at that time, because already conforming to classical electrodynamics, wherein a radiating electron was assimilated to an oscillator, such a radiation should have had to be taken into consideration whenever the electrons have an accelerated motion, not only when they are braked one way or other, but also when their velocity increases. Actually just for this already acknowledged radiation emitted by electrons in accelerated motion, Bohr was obliged some years before to begin his atomic theory with a postulate about the *stationary* atomic orbits, excepted for unspecified reasons from this general rule. Moreover, even in the Bohr's atom the electron radiates a photon only when its linear velocity increases by passing on an orbit nearer to nucleus. And now the best argument in this regard is undoubtedly a comparison between an accumulating ring for monoenergetic electrons and a synchrotron: in both cases the captive electrons move uniformly in a high vacuum on a circular trajectory perpendicular to the flux lines of a uniform magnetic field, but in a storage ring the electrons have constantly the same linear velocity, while in synchrotron they pass at each period through one or several accelerating cavities placed on the circular circumference and supplied with alternative voltage of high frequency, cavities inside of which their linear velocity increases by electromagnetic (photonic) acceleration in a straight line. As a result, if in an accumulating ring the electrons do not radiate, except a weak radiation owed to their incidental collisions with residual gas molecules, in synchrotron they emit during each electromagnetic acceleration inside an electromagnetic cavity a specific photon radiation very much studied experimentally, whose wavelength, emission angle and degree of circular polarization change as the linear velocity of the radiating electrons increases.

It is therefore beyond doubt that this synchrotron radiation is emitted by electrons just because of their electromagnetic accelerations, confirming thus, once again, a phenomenon known for long ago, and more recently noticed in all electromagnetic accelerators, both linear and cyclic: any elementary process of electromagnetic acceleration is accompanied by an own photon emission of the accelerated particles.

But here an unprecedented question appears: should this new term E_γ be included only in the energy equation of the limit frequency method, or is it also obligatory in the Millikan's energy equation? In other words, do the photoelectrons radiate outside during their acceleration from their initial relative rest state to a certain velocity v ? Well, even if until now a photon emission of the photoelectrons during their escape from atomic lattices has not been considered yet, the answer can only be positive. Both the electrons accelerated in electromagnetic fields generated by oscillating currents of high frequency, as it happens in electronic guns or synchrotrons, and the photoelectrons first time pulled out from the atoms, and only then accelerated from the rest to a determinate velocity, convert integrally into their kinetic energy the energy $h\nu$ of the same carriers of electromagnetic interaction at distance, the photons, and the already known energy equivalences $eV = h\nu = m_0 v^2/2$ proves explicitly this similitude between two ways of electromagnetic interaction only artificially differentiated by some. Obviously, the photoelectrons are particles electromagnetically accelerated, even if a part of the energy $h\nu$ of the incident photons is consumed at first for their extraction from an atomic lattice, and only what exceeds their binding energy in atoms plus the extraction work is then converted into their kinetic energy reached finally in the free state. Therefore, there is no doubt that the photon energy E_γ radiated outside by the electrons during their acceleration from the rest state to a certain velocity v has to be unconditionally included in the energy balances of both experiments carried out in the 1910's for measuring the Planck constant.

At the beginning let us introduce this forgotten quantity of energy E_γ into the energy balance of the limit frequency method. In such an experiment eV is the only energy received by the electrons from the outside, while E_γ and E_{br} are the two energy quantities radiated outside by them in the time interval between their initial and final rest state, first of them during their positive acceleration, the second during their subsequent deceleration. Consequently, according to the law of energy conservation we have on aggregate a new energy balance of this experiment

$$eV = E_\gamma + E_{br},$$

different from that used by Blake and Duane, $eV = E_{br}$. And because, on the other hand, we already have the Thomson's equality $eV = m_0 v^2/2$, it results $E_\gamma + E_{br} = m_0 v^2/2$, therefore the sum of the two energy quantities radiated outside by an electron during the whole experiment is equal to the kinetic energy $E_k = m_0 v^2/2$ reached by the electron at the end of its electromagnetic positive acceleration. Unfortunately in this complete energy balance $eV = E_\gamma + E_{br}$ of the limit frequency method, in which the energy E_γ radiated by the electrons during their electromagnetic acceleration is not ignored any longer, the problem is just this new term E_γ, since even now there is no formula for estimating this always ignored energy, in spite of its research in synchrotron for about half a century, where both the increase in velocity of the accelerated particles at each of their passing through accelerating cavities and the frequency of photon radiation emitted by them just because of this acceleration could easy be related to each other.

However, if we have in view that the two radiated energies E_γ and E_{br} correspond to photon emissions caused by equal variations of kinetic energy $\Delta E_k = m_0 v^2/2$, during of which the electrons have first an increase of mass $\Delta m = m_0 v^2/2c^2$ and then the same diminution of mass, we can presume for symmetry reasons their equality $E_\gamma = E_{br}$, even if E_γ is radiated by an electron during its positive acceleration from the rest state up to the speed v, while E_{br} during its subsequent deceleration from the final speed v back to the rest state. In consequence, as $eV = E_k = m_0 v^2/2$, $E_\gamma + E_{br} = E_k = m_0 v^2/2$ and $E_\gamma = E_{br}$, the complete energy balance of this method including beside the elementary electric charge e only the experimentally measured quantities V and ν_{max}, becomes

$$eV = 2h\nu_{max}.$$

Evidently this corrected energy balance means an exactly halved value $h = 3.312 \cdot 10^{-34}$ J·s of the Planck's constant resulted in 1917 from experimental data of the limit frequency method.

As for the Millikan's method, since the photoelectrons are also electrons accelerated to a certain velocity v on account of the energy $h\nu$ of the field particles carrying electromagnetic interaction and called photons, we have not only $h\nu = m_0 v^2/2 = E_\gamma + E_{br}$, but also $eV_{br} = E_\gamma + E_{br} + E_k$, because the braking electromagnetic field has to cover not only the two photon energies E_γ and E_{br} radiated by electrons in the surrounding space, but also the work necessary to annul the kinetic energy $E_k = m_0 v^2/2$ of the moving electrons when they have to be brought back in the rest state (in the method of limit frequency the kinetic energy of the moving electrons is not annulled by a braking electromagnetic field whose work should to be had in view in the total energy balance, in that case their kinetic energy is only gradually transferred to the more or less bound electrons casually encountered by them in the atomic lattice where they are kept in).

Accordingly, the correct energy balance of the Millikan's method, which includes besides the known elementary electric charge e only the two experimentally measured quantities ν and V_{br}, becomes

$$eV_{br} = 2h\nu,$$

and thus the real numerical value of the Planck's constant measured by this method was in fact $h = 3.274 \cdot 10^{-34}$ J·s.

Actually the previous equality $E_\gamma = E_{br}$ is not just a logical presumption, it can be deduced starting from a simple observation: the two new energy equations, $eV_{br} = m_0 v^2/2 + E_\gamma + E_{br}$ for the Millikan's experiment and $eV = E_\gamma + E_{br}$ for the other, should give the same value of the Planck's constant if they are really correct, and this is obligatory even when we have the same velocity v in both experiments. Therefore, in this peculiar case we have $eV_{br} = m_0 v^2/2 + m_0 v^2/2 = 2h\nu$ in the Millikan's experiment, just because $eV = m_0v^2/2 = E_\gamma + E_{br}$ in the other method, and thus the energy balance of the latter becomes $eV = 2h\nu_{max}$, which means $E_\gamma = E_{br}$.

As one can see, when all photon emissions caused by a change in the linear velocity of the used electrons are taken into account, in both energy equations of the two reexamined experiments a new factor 2 appears before the constant h, and these two identical corrections mean a Planck's constant twice smaller than the one accepted today.

However a question becomes unavoidable now: how could such a monumental error remain unnoticed in theoretical and experimental research in the next decades, even in those experiments — very few, indeed ! — where the energy of some monochromatic photons with known frequency was indirectly measured?

Well, the response is surely amazing for many: this major discrepancy between the Planck's constant resulted from the two experimental measurements in the 1910's and the real value of this fundamental constant appeared clearly in the next period, at least in the two cases known as magnetic resonance and Josephson transitions, but each time it was preferred to dissimulate it through ad hoc theories, instead of a thorough reexamination of the two former experiments whose energy equations were obviously incomplete even in those early times of particle physics.

Electronic magnetic resonance

In 1930 Appleton and Childs[3] imagined a device able to measure very accurately the electron magnetic moment M_e, whose size was already estimated in 1921 by Stern and Gerlach, but not with high accuracy. In their experiments Appleton and Childs sent a linear beam of electrons on a direction parallel to the flux lines of a uniform magnetic field $\bar{B}$, while an oscillating current of variable frequency ν generated a pulsating electromagnetic (photonic[4]) field perpendicular to the field $\bar{B}$. When the energy variation ΔE of the electrons after their entering the uniform magnetic field is equal to

---

[3] E. V. Appleton and E. C. Childs, *On some radio-frequency properties of ionized air*, Phil. Mag. **10**, (1930) 969.

[4] And not a pulsating *magnetic* field, as falsely it is claimed sometimes by an interested confusion between an electromagnetic field, which acts on electrons only when they absorb its field particles (the photons), and a magnetic field, which acts on all electrons within its effective radius of action.

the energy $h\nu$ of the radiated photons with motion directions perpendicular to $\bar{B}$, $\Delta E = h\nu$, a resonance phenomenon is detected.

Later on, this method was called magnetic resonance and has been successfully used by many other researchers for measuring with high precision the magnetic moments of the atoms, light nuclei or elementary particles, mostly those having nuclear magnetism.

The clue of this method is the energy variation ΔE of an electron (or any other magnetic particle) after its entering the uniform magnetic field $\bar{B}$. In the 1930's it was already known that the fine or hyperfine splitting of the spectral terms appears because in atoms the radiating electrons have their magnetic moment $\bar{M}_e$ orientated whether parallel or antiparallel to some intra-atomic magnetic fields, and the energy gap $\Delta E = 2M_e B$ between these two possible orientations does not depend on the velocity and the energy of the electrons, but only on magnetic induction B experienced by them in each allowed atomic state. Moreover, these two exclusive orientations have also been adopted in quantum theory both for the electron and proton[5], even if the quantum rules refer explicitly to their spin.

Accordingly, as the energy E of the electrons before entering the uniform magnetic field $\bar{B}$ is evidently placed at the middle of the energy gap $\Delta E = 2M_e B$ delimited by the minimum energy $(E - M_e B)$ corresponding to parallel orientation $\bar{M}_e \uparrow\uparrow \bar{B}$ and the maximum energy $(E + M_e B)$ corresponding to antiparallel orientation $\bar{M}_e \uparrow\downarrow \bar{B}$, it is also beyond doubt that the energy variation experienced by the electrons when they enter the uniform magnetic field $\bar{B}$ is always

$$\Delta E = M_e B,$$

no matter if inside the field their magnetic moments are parallel or antiparallel to the field direction.

Undoubtedly all these experiments of magnetic resonance ought to have been solved from the very outset in accordance with these clear provisions of the basic laws of electromagnetism, but experimental data pointed out an unexpected and very inconvenient disagreement: the correct resonance $\Delta E = M_e B = h\nu$ given an action constant $h = 3.313 \cdot 10^{-34}$ J·s twice smaller than the one considered till then, therefore exactly that one resulted

---

[5] M. Gell-Mann and E. P. Rosenbaum, *Elementary particles*, Sc. Amer. **197** (1), (1957) 72.

when the flagrant omissions in the two energy equations used by Millikan and Duane are at last considered!

But because such an upheaval could not be accepted by the ones vitally interested in saving their recently accepted theories from a not too glorious denial, it became imperatively necessary a reason for doubling the energy variation $\Delta E = MB$ felt by magnetic particles with magnetic moment M inside a uniform magnetic field B, and the found solution was the older theory of the spin precession around the flux lines of the external magnetic field (originally based on an analogy between the electron spin and gyroscope, although in quantum physics is always emphasized that the electron spin is an intrinsic property of the particle, absolutely unrelated to a rotational motion of it).

Indeed, this theory presumes also a possible reversal of the spin direction when an electron within a uniform magnetic field $\bar{B}$ absorbs a photon whose frequency is equal to that of the spin precession motion, and this flip-flop of the two antiparallel moments of the electron, angular $\bar{L}_e$ and magnetic $\bar{M}_e$, implies evidently an energy variation $\Delta E = 2M_e B$ of the electron, which has been adopted as the saving equation of magnetic resonance, valid for all magnetic particles having magnetic moment M,

$$\Delta E = 2MB .$$

Or, this insertion of the spin precession theory in magnetic resonance has ignored an elementary question: the spin precession motion was postulated for atomic electrons in orbital motion, because such a spin precession around the external magnetic field is possible only by a simultaneous existence of both orbital moments of the involved particles, kinetic and magnetic, this is an absolutely obligatory condition in the theory of spin precession in magnetic field. But such orbital moments cannot appear in magnetic resonance experiments, simply because the Lorentz force exerted on the particles in inertial motion along the parallel magnetic flux lines is null, which excludes even their motion on curved trajectories, and all the more a genuine orbital motion of them.

Therefore, the energy gap $\Delta E = MB$ between the two different energies that any motionless or moving elementary particle with magnetic moment M has outside and, respectively, inside a uniform magnetic field B was entirely ignored, as this energy difference predicted by the laws of electromagnetism and experimentally noticed in all line spectra simply would not exist, and interestedly replaced by the double variation of energy $\Delta E = 2MB$ valid for spin inversions in uniform magnetic field, although in these experiments of magnetic resonance such spin inversions cannot occur as long as the involved magnetic particles have only an inertial, rectilinear motion along the field lines.

There is, however, an experimental *detail* whose reconsideration now becomes a strong argument in favor of the above: actually, Appleton and Childs found in their experiments with electrons not a single maximum of resonance, but two maxima, that of higher amplitude leading to the final result $M_e = M_B$, and that of smaller amplitude to the twofold value of the electron magnetic moment $M_e = 2M_B$! And this second maximum of resonance for a twice smaller wavelength of the photons emitted on directions perpendicular to $\bar{B}$ was also noticed by other researchers[6], but remained then with no reliable explanation and later forgotten.

Or, now just this ratio 2/1 between the two resonance wavelengths proves their only possible origin: while the first and stronger maximum of resonance can only correspond to the energy variation $\Delta E = M_e B$ instantaneously experienced by all electrons when they enter the uniform magnetic field B, the second maximum of lower intensity can only appear because a certain fraction of the already entered electrons really experiences a subsequent spin inversion in the uniform magnetic field B, either because their motion directions are inherently a little divergent from the parallel force lines of the field $\bar{B}$, and thus the electrons experience yet a weak Lorentz force and have yet a slightly curved trajectory resulted from the superposition of a cyclotron motion to their inertial motion along the force lines of the uniform magnetic field, or such spin inversions in uniform magnetic field simply have causes entirely out of this postulated theory of spin precession.

Anyway, it remains evident that the true equation of magnetic resonance is that corresponding to the main maximum $\Delta E = MB$, caused by the doubtless energy variation experienced by all the incident electrons right at the moment of their entry into the uniform magnetic field B, while the secondary maximum corresponding to the energy equation $\Delta E = 2M_e B$ reveals a subsequent spin inversion performed only by a part of the incident electrons.

---

[6] Th. V. Ionescu and C. Mihul, *Les gaz ionisés dans le champ magnétique; preuve de l'existence de l'électron tournant*, C. R. Acad. Sci. Ser. B **194**, (1932) 70.

And such an absolutely interested arbitrary insert of a 2-factor in an energy equation including the Planck's constant repeated for similar reasons after about three decades, this time owing to some surprising experimental data concerning electronic transitions in superconducting metals, which proved again in vain that something is wrong regarding the always accepted value of the Planck's constant.

Josephson transitions

In 1968 Clarke[7] measured exactly both the chemical potential differences V corresponding to the induced steps of some electronic transitions in different superconducting alloys (containing Sn, Pb, In) and the frequencies ν of the photons emitted during these transitions, a perfect way for measuring the Planck's constant as a proportionality factor between the energy eV and the frequency ν of that transition photons, $eV = h\nu$. But because this normal energy equation led again to a troublesome value $h = 3.313 \cdot 10^{-34}$ J·s of the Planck's constant, the energy equation of these experiments was again deliberately distorted, this time by doubling the number of the electrons involved in each elementary electronic transition.

In short, these transitions have been considered to be performed not by single electrons, as it happens in all the other electronic transitions, but by some *virtual*[8] pairs of electrons imagined some years before in a new quantum theory of superconductivity[9], pairs made of electrons bound together by the action of another *virtual* particle called phonon (actually a vibration of the whole atomic lattice inside of which these pairs of electrons occur!) able to overcome the repulsive electrostatic interaction between the two paired electrons, so that the energy of the single photon emitted during their transition from a higher level of energy to another with smaller energy is twice higher than the energy difference between the two involved energy

---

[7] J. Clarke, *Experimental comparison of the Josephson voltage-frequency relation in different superconductors*, Phys. Rev. Lett. **21**, (1968) 1566.

[8] Imaginary, something existing only as a possibility, not in fact.
Existing in the mind, especially as a product of the imagination.
Very close to being something without actually being it.

[9] J. Bardeen, L. N. Cooper, and J. R. Schrieffer, *Microscopic Theory of Superconductivity*, Phys. Rev. **106**, (1957) 162.

levels, and therefore the energy equation of these electronic transitions gets the absolutely unprecedented form[10]

$$2eV = h\nu.$$

Or, the explicit meaning of this visibly and interestedly *well-arranged* equation is exactly an emission *in common* of only one photon by two electrons presumed to be paired by an ad hoc fabricated particle, but two electrons distinct yet!, as long as $2eV$ is evidently the sum of the two identical quantities of energy eV lost separately by each of the two electrons, even if they would be bound to each other one way or other, while $h\nu$ is the energy of a single miraculous photon able to take over both quantities of energy lost individually by two distinct electrons during their simultaneous jump on a lower level of energy!

That a single photon can be shared out by two distinct electrons is proved experimentally when one and the same photon excites simultaneously two atoms in touch, providing its energy $h\nu$ to be equal to the two added excitation energies of the atoms, no matter if the latter are equal or not[11], but an emission in common of a single photon by two electrons even intimately bound, but yet distinct !, is something entirely nonsensical, an evident and unacceptable logical fracture. Obviously, if a pair of electrons pass together from a certain state into another of smaller energy, each of these two distinct electrons has to emit its own transition photon in a random direction, and thus each transition of this kind has to give rise to a pair of photons with the same frequency, but different propagating directions. In consequence the energy equation of the so-called Josephson transitions remains indisputably that one valid for all electronic transitions between two discrete levels of energy, $eV = h\nu$, and thus the true numerical value of the Planck's constant resulted from experimental data of Josephson transitions is once more $h = 3.313 \cdot 10^{-34}$ J $\cdot$ s.

The two deliberately and grossly distorted equations, $2MB = h\nu$ for magnetic resonance and $2eV = h\nu$ for electronic transitions in metal lattices, both of them by introducing a saving 2-factor no matter how, even by infringing the elementary logic, show up how far the despair can lead. And

---

[10] B. D. Josephson, *Possible new effects in superconductive tunneling*, Phys. Lett. **1**, (1962) 251.

[11] M. H. Mittleman, *Excitation of two atoms by a single photon*, Phys. Lett. A **26**, (1986) 612.

that despair had to be really great, if both quantum theories fabricated for justifying these ineptitudes were rewarded with Nobel prizes[12] !

No other self-consistent measurements of action constant

Previously critical reexamination of the four most reliable ways for measuring the action constant reveals a reality hard to be imagined for a quantum physics seen in the modern age as an acme of the human knowledge: after more than a century of theoretical research in quantum physics it appears the unexpected prospect of rebuilding from scratch the whole physics of microcosm owing to only one essential *detail*, a wrong value of "the Planck's constant, h, a universal constant that lies at the heart of quantum mechanics"!

Fortunately, all the circumstances exclude a precipitate sentence in this matter:

(1) Both experiments carried out in the 1910's for measuring the Planck's constant have had evidently incomplete energy equations, which have ignored the photon emissions outside of the electrons during their electromagnetic acceleration and deceleration, photon emissions whose reality was already acknowledged even in those times. And when these unacceptable omissions are at last corrected, both experiments give a new value $h = 3.313 \cdot 10^{-34}$ J·s for the Planck's constant, twice smaller than that one always considered in theoretical physics;

(2) Ulterior experiments, as magnetic resonance and Josephson transitions, have had energy equations deliberately fitted to correspond to the needed value $h = 6.626 \cdot 10^{-34}$ J·s of the Planck's constant, but in both cases their adaptation by doubling the energy of the involved photons infringes flagrantly the basic laws of physics, and even of logics. Without that supplementary 2-factor interestedly introduced, both their energy equations confirm a halved value $h = 3.313 \cdot 10^{-34}$ J·s of the action constant. There is, therefore, a general convergence of the arguments for a halved numerical value $h = 3.313 \cdot 10^{-34}$ J·s of the Planck's constant, which precludes any doubt on its correctness.

---

[12] F. Bloch and E. M. Purcell in 1952 for the interestedly false equation $2MB = h\nu$ of magnetic resonance, B. D. Josephson in 1973 for the absurd equation $2eV = h\nu$ of some electronic transitions in atoms.

It must also be specified that the four previously reexamined experiments — Millikan, limit frequency of X-ray bremsstrahlung, magnetic resonance and Clarke — have queerly remained until now the only able to measure the Planck's constant in the most direct and self-consistent manner, as a proportionality factor $h = E_\gamma/\nu$ between the frequency ν and the energy E_γ of some monochromatic photons, evidently providing that their energy equations to not be altered by omissions or interested distortions. All the other methods subsequently proposed have not been self-consistent any longer, for example the methods called Watt balance and Faraday constant use the same nonsensical energy equation $2eV = h\nu$ interestedly invented by Josephson, while the method based on X-ray diffraction uses a Bragg's equation including again a supernumerary 2-factor, this time resulted from a geometrically unjustified doubling of the path difference caused when the incident X-rays are reflected on two neighboring, parallel reticular planes in a crystal.

In fact, all these more recent methods have been imagined just for getting as much as possible accurate significant digits in an already preestablished Planck's constant $h = 6.626 \cdot 10^{-34}$ J·s . Instead, in spite of all issues raised a long time ago by experiments as magnetic resonance or Josephson transitions, no one has considered until now some very simple and direct methods for verifying if the numerical value always used in theoretical physics is really beyond doubt, for instance by using the Clark's method in other electronic transitions of lower frequency, where the Cooper pairs cannot be invoked, or by measuring the speed of the fastest photoelectrons emitted when a beam of monochromatic light with known frequency enter a metal lattice rich in quasi-free electrons roaming among the ionized atoms, photoelectrons entirely similar to those used by Millikan in 1914.

Indeed, when such an atomic lattice is lit, say, by a beam of red light with known frequency $\nu = 4 \cdot 10^{14}$ Hz , the highest velocity $v = \sqrt{2h\nu/m_0}$ of resulting photoelectrons is equal to $7.6 \cdot 10^5$ m/s if $h = 6.626 \cdot 10^{-34}$ J·s , but to only $5.4 \cdot 10^5$ m/s if $h = 3.313 \cdot 10^{-34}$ J·s . And because in the first case the fastest photoelectrons emitted by the incident red light would cover in vacuum a distance equal to 1 m in $1.31 \cdot 10^{-6}$ s , but in $1.85 \cdot 10^{-6}$ s in the second case, by using an equipment able to measure the time intervals with a maximum error 10^{-9} s , already available for many decades, the real value of the Planck's constant would have been definitively established also long ago. Is there an acceptable explanation for this strange lack of interest in an experiment so simple, but so relevant?

Besides such measurements absolutely necessary just because no one could question the significance of their results any longer, a crossing remake of the two experiments[1,2] performed in the 1910's would undoubtedly be very spectacular:

(1) An experiment with electrons first time accelerated from the their relative rest state to a certain velocity in an electromagnetic field V, and then brought back at rest state in a delaying electromagnetic field V_{br};

(2) Determining the highest frequency ν_{max} of the bremsstrahlung evolved in an atomic lattice by monoenergetic photoelectrons emitted with negligible extraction work from a metal lattice rich in quasi-free electrons by monochromatic light with known frequency ν.

In truth, if conforming to the physicists in the 1910's their results should be

$$V_{br} = V \quad \text{and} \quad \nu_{max} = \nu,$$

when the two photon emissions E_γ and E_{br} ignored by them are taken into account, the predicted ratios are entirely different,

$$V_{br} = 2V \quad \text{and} \quad \nu_{max} = \nu/2,$$

so that the true value of the Planck's constant could be established beyond doubt.

And such measurements are certainly quite feasible with the current technical means!

Return to spinning ring electron

Now the Planck's constant is the proportionality factor between the frequency ν and the energy E_γ of the photons, $E = h\nu$, but it is also the natural unit for all spins of elementary particles or orbital kinetic moments in the atoms.

First of these two primordial attributes have an overwhelming importance in physics, since the photons have always been one of the main sources of information about what happens in the near microcosm. Their prominent part in this respect was determined by an early development of spectroscopy, by means of which their frequencies could be accurately measured, from those of microwaves, which are the photons with the smallest frequency in nature, to the upper limit about 10^{19} Hz of the hard X-rays. But because in all interactions involving photons what matters above all is the energy of the latter, equal to the energy lost by their elementary emitters or acquired by their receivers, the frequency scale of electromagnetic radiation has been doubled by an energetic scale whose values are determined by multiplying the measured frequencies ν by the action constant h, $E = h\nu$. Only at the upper end of the two parallel scales the procedure has been inverted under constraint: because the photon frequencies over about 10^{19} Hz can be directly measured by no method, it has been determined by specific methods the energy E of certain monochromatic photons which can reach even over 10^2 MeV, and then their frequency is calculated by the same universal relation, $\nu = E/h$. Still this change of method in the γ-rays range does not affect the continuity of the two scales, no hiatus appears when the frequency measurements are replaced by energy estimations, so that the action constant h remains the proportionality factor between the frequency values ν and their corresponding energy values $E = h\nu$ on the entire length of the two parallel scales of electromagnetic radiation.

In these circumstances, the energetic scale of electromagnetic radiation is essentially affected by halving the action constant h, because for the same frequency scale, whose values ν are experimentally measured, all corresponding values $E = h\nu$ on the parallel scale of energy become automatically twice smaller than until now.

Or, it is not at all hard to understand that a halving of all values on the whole energetic scale of electromagnetic radiation, from the microwaves to the hardest γ-rays, means a sudden collapse for the most of fundamental theories elaborated until now in physics of microcosm. And the most spectacular case is probably the halving of the energy about 0.5 MeV found experimentally by Klemperer[13] in 1934 for the two identical γ-photons resulted from annihilation $e^- + e^+ \to 2\gamma$ at rest, energy evidently equal to the rest energy E_0 of the electron.

But if the energy of each of the two annihilation photons is in fact equal to only ~ 0.25 MeV, and therefore the formula of the rest energy of the elec-

---

[13] O. Klemperer, *On the annihilation radiation of the positron*, Math. Proc. Cambridge Phil.Soc. **30**, (1934) 347.

tron becomes $E_0 = m_0 c^2/2$, we have a clear conclusion whose denial is quite impossible: the electron cannot be a material point with no inner structure, as it is considered in quantum relativistic physics, because the new rest energy $E_0 = m_0 c^2/2$ is obviously an inner kinetic energy of a composite electron at rest, whose structure can only be that one of a rotating ring with a linear peripheral velocity equal to c. And because the relativistic rest energy $E_0 = m_0 c^2$ of the electron means quantum relativistic physics, while the internal kinetic energy $E_0 = m_0 c^2/2$ of the electron means classical physics, we can understand that the previously demonstrated compulsoriness for halving the current Planck's constant means implicitly an obligatory, ultimate return to the classical model of spinning ring electron.

More, we should to not forgot the key position of the electron in the table of all the known elementary particles and their mutual transformations: indeed, the electrons emit or absorb photons and neutrinos, they annihilate into photons and in their turn can be created from photons, the electrons represent the final stable state of the unstable muons and charged pions, the nucleons and hyperons can be synthesized from electrons of high energy, and inversely the unstable nucleons and hyperons emit electrons and positrons, and so on.

In consequence, once a believable structure of the spinning ring electron is found and confirmed by its capacity to explain the known properties of this first elementary particle, this really structural model of the electron paves by itself the way for deciphering the inner structure of all the other main elementary particles, and even the deep-seated mechanisms of their creation or disappearance. Nothing is more evident than the difference between the research potential of this live spinning ring electron and that of the inert pointlike electron of quantum relativistic physics, with no inner structure but with *intrinsic* (a merely empty word!) rest energy, spin and magnetic moment.

But not only the theory of relativity disappears by halving the numerical value of the Planck's constant, all atomic theories are also demolished by a halved action constant, because all their quantizing formulas are evidently invalidated by halving both the energy levels and the spin or orbital kinetic moments of atomic electrons.

If a collapse of theoretical works built up on a Planck's constant twice higher than its real value is undoubtedly a normal consequence when this overvaluation is proved, problems appear in understanding how theoretical constructions wrong right from the start, as now quantum atomic physics and special relativity appear, could still withstand so long time, or, in other words, why no experimental data have revealed their falsity yet. Indeed, in this regard no serious trouble can be found in the current courses and treatises, on the contrary, they compete exclusively for proposing more and more experimental confirmations for these theoretical constructions. And then, are not quantum atomic physics and theory of relativity by themselves multiple experimental confirmations of the always acknowledged Planck's constant $h = 6.626 \cdot 10^{-34}$ J·s, used so long time by them?

At first sight the answer could only be affirmative, but when some want verifying some elementary aspects strangely never noticed and discussed in specialized literature, for example how experimental data regarding the fine and hyperfine splitting in hydrogen line spectrum confirm the primary quantizing equation $mvr = n\hbar$ postulated in semi-quantum or quantum theories of hydrogen atom, or how special relativity explains synchrotron radiation, the things change in all.

And because these two examples are not at all singular in this regard, the next two chapters of this book are allotted for presenting other telling aspects in atomic physics and theory of relativity, which can be found hardly ever in the courses or treatises in these branches of physics. Thus any reader of today can judge for yourself how rightful both atomic physics and theory of relativity have been to replace the spinning ring electron adopted by experimental researchers with their pointlike electron destitute of inner structure, but with inner energy, magnetic moment and angular momentum. That is why these two next chapters about atomic physics and theory of relativity should not be seen as an end in itself, but only as complementary arguments against the quantum and relativistic pointlike electron, and implicitly for a necessary reinstatement of the old classical model of the spinning ring electron.

NOTIONAL ATOMIC HYDROGEN

Continuous end of spectral series assigned to atomic hydrogen

The line spectrum assigned to atomic hydrogen is made of all allowed energy levels of the electrons in the hydrogen atoms presumed to result by dissociating molecular hydrogen passed through a discharge tube. Each of these discrete energy levels, or spectral terms, corresponds to an allowed atomic state of the free hydrogen atoms, and all these spectral terms form together the so called Grotrian diagram of atomic hydrogen, limited at the lower end by the fundamental term $1S$ representing the energy level of the unexcited atom, and at the other end by a continuous area, which includes all unquantified energy levels of the free negative photoelectrons just removed from hydrogen atoms by photoionization and ready for a new binding to the free positive protons because of their electrostatic attraction.

But here all atomic theories, from Bohr's semi-quantum theory to modern quantum theory of hydrogen atom, raise an insuperable problem, because a moving electron and a moving proton can form a stable two-body system in dynamical equilibrium only if their initial velocity vectors are perfectly correlated, both in absolute values and as directions, and such a perfect correlation has practically a null probability. For this reason, when a slow free proton and a slow free electron get close to one another, their electrostatic attraction, which is a central force permanently placed on the line joining them, can lead whether to their final collision after a longer or shorter trajectory depending on their initial velocity vectors, in which case it should occur a reaction $p^+ + e^- \to n^0 + \gamma$, as it happens in electronic captures in atoms, or only to a mutual deviation of their trajectories, but not at all to their binding in a planetary system proton-electron. And in the simplest case, that of a proton and an electron initially at rest at a very great distance from each other, the two particles can only fall toward each other on the line joining them until their final collision. For all that, despite the evidence, even today the lowest limit of the continuous zone in the hydrogen line spectrum is assigned to an impossible coupling of the free protons and electrons initially at rest and the formation of the excited or unexcited hydrogen atoms!

As a matter of fact, in the course of time all experimental attempts to obtain hydrogen from colliding beams of very slow free protons and electrons failed with no exception.

It is very clear, forming the hydrogen atoms from free moving protons and electrons is only a nonsense thought in a primitive age of physics of microcosm, an idea in total disagreement with the basic laws of mechanics. Therefore, that terminal continuous area in the line spectrum of hydrogen disproves right from the start all atomic theories developed in the course of time just because all of them explain it by a physically impossible process of forming the free hydrogen atoms through electrostatic attractions between positive protons and negative electrons.

But the irreconcilable conflict between atomic theories and the basic laws of physics does not stop here.

Fine and hyperfine splitting in hydrogen line spectrum

Soon after the appearance in 1914 of the Bohr's atomic theory, a rapid progress of spectroscopy led first time to the discovery of the fine structure of hydrogen spectral lines, and then of their hyperfine structure. Splitting of the spectral lines was already know since 1886, when Zeeman recorded split spectral lines emitted by hydrogen in uniform magnetic fields of different strengths. And indeed, these structures proved to be consequences of the fine splitting of P, D and F spectral terms in hydrogen line spectrum, respectively of the hyperfine splitting not only of S terms (which have not fine splitting), but also of all P, D and F fine subterms, all of them being of magnetic nature. The energy gap ΔE of any spectral split in magnetic field B is always given by the formula $\Delta E = 2M_e B$, so that it depends exclusively on magnetic induction experienced by the radiating electron during its photon emission, and not on its speed or other physical parameter.

Therefore, such simultaneous fine and hyperfine splits of the energy levels in hydrogen spectrum assigned to the free H atoms can only appear if the radiating atomic electrons emit their transition photons within two distinct intraatomic magnetic fields, the stronger responsible for the fine splitting effect, the weaker for that hyperfine. For example, while the fine splitting of the $2P$ term is equal to 10,969.1 MHz, the hyperfine splitting of its fine subterm $2P_{1/2}$ is only 59.19 MHz [14], and this large difference proves a similar disproportion between the strengths of the two magnetic fields responsible for these distinct energy splits (usually all the fine or hyperfine energy splits ΔE are expressed as frequency gaps $\Delta v = \Delta E/h = 2M_e B/h$).

Or, as in his theory the atomic origin of the hydrogen line spectrum was implied right from the start, since the predominant presence of the free atoms of hydrogen in rarefied gaseous hydrogen previously passed through vacuumed discharge tubes was already an older belief in those days, Bohr had also to find the sources of the two intra-atomic magnetic fields responsible for the fine and hyperfine structures of the spectral lines emitted by these simplest atoms, which contain besides the radiating electron only the nucleus-proton. And if the weaker hyperfine effect could be caused only by the dipole magnetic field generated by nucleus-proton, the other necessary intra-atomic magnetic field, responsible for the fine splitting of the P, D and F spectral terms and therefore of hundreds times stronger than the proton magnetic field, seemed some years to be a problem definitively insoluble for Bohr's atomic theory, able even to disprove it in all. However, at the Pauli's suggestion, Goudsmit and Uhlenbeck *resolved* this very annoying difficulty in 1925, when they still were undergraduates in physics, by identifying as responsible for this stronger fine effect the dipole magnetic field generated in any excited hydrogen atom by ... the orbital motion of the radiating electron itself! And this monumental nonsense, unacceptable even for undergraduates, was at once accepted by Bohr, and then assumed in all treatises of atomic physics, even until today!

In reality, as it is clearly demonstrated in any elementary treatise of electromagnetism, in accordance with the Biot-Savart law a pointlike electron in motion generates a magnetic field whose force lines are every moment circles concentric to its motion direction, and the subsequent spatial configurations of this constantly changing field are not cumulative in time, no matter how rapid is the motion velocity of its elementary source[15], while a

---

[14] S. J. Brodsky and D. Drell, *The Present Status of Quantum Electrodynamics*, Annual Review of Nuclear Science, edited by E. Segré, Palo Alto, California, 1970, Fig. 12.

[15] If such a persistence in time would really exists, then even a very fast electron in inertial motion ought to originate around its propagating direction a magnetic field identical to that generated by a long linear fascicle of very close electrons, or, at macroscopic level, by a linear stationary electric current. And such hilarious examples could continue.

dipole magnetic field defined by a magnetic moment can only be generated by a plane closed current whose electric charges are uniformly spread on its whole trajectory, as a vector sum of all Biot-Savart magnetic fields generated simultaneously any minute by each elementary electric charge in part.

More, to say about a dipole magnetic field generated by the orbital motion of an electricity quantum and able to act even on its elementary source oversteps any limit of a sick inventiveness.

And then, why is such a grossly false accepted even today?

Very simple, because in hydrogen atom does not exist a second electron whose dipole magnetic field could have justified the fine splitting of the energy levels P, D and F of the radiating electron, a second electron exists only in hydrogen molecule, but who could call into question such a variant when Bohr's theory of atom just was rewarded with Nobel prize? But what is more interesting, the atomic origin of the hydrogen spectral lines could be easily verified later with the aid of the main available experimental data[14] referring to the fine and hyperfine splitting $\Delta\nu = 2M_e B/h$ in this spectrum.

Such verification is very simple if two aspects are considered:

(1) In all atomic states of the Bohr's atom, the total energy E of atomic electrons is inversely proportional to their orbital radii r, $E = e/2r$;

(2) The dipole magnetic field with axial symmetry of the central proton in hydrogen atom has to be identical to those of all the other magnetic particles in microcosm (electrons, neutrons, muons, charged pions, etc.), and ultimately to those generated at macroscopic level by the circular electric currents, whose known equation[16] is

$$B = \sqrt{(1 + 3\cos^2\theta)}\, \mu_0 M /4\pi r^3,$$

where B is magnetic induction in a certain point in the field, M is the moment of the dipole magnetic field, $\mu_0 = 4\pi \cdot 10^{-7}$ H/m is magnetic permeability of vacuum, while $\theta = 0 \ldots 2\pi$ is the angle between the symmetry axis Ox of the field and the vector radius $\vec{r}$ of the considered point.

Who is still not sure enough in this regard should take into account that *magnetic moment* is a physical quantity proper only to the dipole magnetic

---

[16] E. M. Purcell, *Electricity and Magnetism*, Berkeley Physics Course, Vol. II, Mc Graw-Hill, New York, 1965.

fields generated by closed and plane stationary electric currents, which is clearly stated by its defining equation itself, $M = IS$, where I is the constant intensity of a closed electric current sited in a plane, and S is the area delimited by its circumference.

Therefore, even the dipole magnetic fields of the elementary particles can be generated only by plane closed currents existing in their subquatum space, another physical explanation for their noticeable and measurable magnetic moments can exist just in the magic world of physics fiction. And because these dipole magnetic fields exhibit always an axial symmetry, their subquantum close and plane currents can only be circular.

Actually this simple understanding of the hyperfine splitting effect was explicitly expressed long ago:

"There is a general and widely prevalent misconception that the hyperfine interaction can be understood only by someone who has first become familiar with the relativistic Dirac equation for the motion of an electron in the Coulomb field of the atomic nucleus. (...) It is therefore desirable to establish a simple conceptual picture of the origin of the hyperfine interaction which does not depend upon something so sophisticated as quantum mechanics, much less the Dirac equation.

(...) The preceding derivation of the hyperfine interaction shows that is possible to understand the origin of this coupling purely from the standpoint of simple magnetic considerations. Quantum mechanics is irrelevant to this interaction, which is of a purely classical nature."[17]

Why quantum physics has always avoided to consider the evident origin of the hyperfine splitting effect becomes very clear when the above formula of all dipole magnetic fields with axial symmetry is applied to the nucleus-proton in the hydrogen atom for verifying if the spectroscopically measured hyperfine splits of hydrogen spectral terms correspond with the very simple structure of this atom.

Right from the beginning we have to admit that the rigorously constant and sharp values of all hyperfine splits measured for hydrogen spectral terms can be explained only if in all allowed states in hydrogen atoms the radiating electrons experience the same nuclear magnetic induction at

---

[17] R. A. Ferrel, *Electron-Nucleus Hyperfine Interaction in Atoms*, Amer. J. Phys. **28**, (1960) 484.

every point on their orbits, which is evidently possible only if all these allowed orbits are perfectly circular. Otherwise, any hyperfine gap in the fine spectral subterms P, D and F would vary depending on the point of the orbit wherefrom the radiating electron emits its transition photon. Or, this condition totally disagrees with the elliptical orbits predicted by quantum mechanics for the p, d and f states of the hydrogen atoms.

The quantitative verifications are also very relevant.

(1) By using the above formula valid for all dipole magnetic fields with axial symmetry, including those generated by elementary particles, we can calculate the orbital radius r_{1s} of the atomic electron in the fundamental state 1s of this atom from the hyperfine splitting $\Delta v_{1S} = 1.420406 \cdot 10^9$ Hz experimentally measured for the fundamental term 1S in the line spectrum assigned to hydrogen atom[14],

$$r_{1s} = \sqrt[3]{2 \cdot 10^{-7} \sqrt{1 + 3\cos^2\theta}\, M_e M_p / h \Delta v_{1S}}\ ,$$

where $M_e = M_B = 9.273 \cdot 10^{-24}$ A·m² is magnetic moment of the atomic electron and $M_p = 1.420 \cdot 10^{-26}$ A·m² is that of the proton. It results the minimum radius $r_{1s_{min}} = 0.303 \cdot 10^{-10}$ m for $\theta = \pi/2$ and the maximum radius $r_{1s_{max}} = 0.382 \cdot 10^{-10}$ m for $\theta = 0$, therefore all orbital radii possible in hydrogen atoms in fundamental state $n = 1$ are significantly smaller than the so-called Bohr radius $r_B = 0.529 \cdot 10^{-10}$ m.

More, these minimum and maximum calculated radii correspond accurately to experimentally measured radii $r \approx 0.3 \cdot 10^{-10}$ m of all hydrogen atoms bound in molecules, excepting those bound in hydrogen molecules, whose experimentally measured radii are $r \approx 0.38 \cdot 10^{-10}$ m. And certainly this double coincidence cannot be accidental.

(2) The hyperfine splitting $\Delta v_{1S} = 1.420406 \cdot 10^9$ Hz of fundamental term 1S in the hydrogen line spectrum proves to be exactly eight times larger than the hyperfine splitting $\Delta v_{2S} = 1.77556 \cdot 10^8$ Hz measured for the 2S term in the same line spectrum[14], $\Delta v_{1S}/\Delta v_{2S} = 8$. But this very exact ratio means another very exact ratio $r_{2s}/r_{1s} = \sqrt[3]{8} = 2$ between the orbital radii of the two atomic states 2s and 1s. Or, this experimental result disproves not only the Bohr's semi-quantum theory, which calculates $r_{2s} = 4\, r_{1s}$, but also the quantum mechanics of hydrogen atom, whose specific probability factor $\Psi\Psi^* dr$ has a maximum for $r_{2s} \approx 6\, r_{1s}$. And when we con-

sider also the known velocity ratio $v_{2s} = v_{1s}/2$ of the atomic electron in the two states $2s$ and $1s$ in hydrogen atom, it becomes as clear as possible that the two spectroscopically measured hyperfine splits Δv_{1S} and Δv_{2S} prove an unchanged orbital moment of the radiating electron in the two atomic states $1s$ and $2s$, $mv_{1s}r_{1s} = mv_{2s}r_{2s}$, therefore a constant orbital kinetic moment of the atomic electron in all atomic states of hydrogen atom,

$$mvr = \hbar,$$

different from the Bohr's postulate assumed by quantum mechanics,

$$mvr = n\hbar,$$

which predicts $mv_{1s}r_{1s} = 2\, mv_{2s}r_{2s}$.

The same denial of the Bohr's atomic theory appears in the Grotrian diagram of deuterium atom, whose homologous terms $1S(D)$ and $2S(D)$ have their homologous hyperfine splits $\Delta v_{1S(D)} = 3.27384 \cdot 10^8$ Hz [18] and $\Delta v_{2S(D)} = 4.09245 \cdot 10^7$ Hz [19] in the same ratio $\Delta v_{1S(D)}/\Delta v_{2S(D)} = 8$, wherefrom it results a ratio of orbital radii $r_{2s(D)}/r_{1s(D)} = \sqrt[3]{8} = 2$ in total disagreement with that predicted by the Bohr's postulate, $r_{2s(D)}/r_{1s(D)} = 4$.

Although these experimental values of the hyperfine splits of the $1S$ and $2S$ terms in hydrogen and deuterium spectra are absolutely enough for disproving definitively the Bohr's quantization equation $mv_n r_n = n\hbar$, the experimentally measured hyperfine splits of the hydrogen and deuterium $3S$ terms would be of great interest in this context, but unfortunately these data are not available in literature. Anyway, in accordance with the new equation $mv_n r_n = \hbar$, the hyperfine splitting of the $3S$ term in the hydrogen line spectrum has to be $\Delta v_{3S(H)} = \Delta v_{1S(H)}/27 = 5.26076 \cdot 10^7$ Hz.

(3) The atomic electron in two atomic states with almost the same energy, therefore with almost the same orbital radius r, but different spatial configurations, therefore with different angles θ, can experience nuclear magnetic inductions B_p differing at most by a 2-factor, corresponding to the maximum value of the angular factor $\sqrt{1 + 3\cos^2\theta}$. As a result, in accor-

---

[18] A. G. Prodell and P. Kusch, *On the hyperfine structure of hydrogen and deuterium*, Phys. Rev. **79**, (1950) 1009 and *The hyperfine structure of hydrogen and deuterium*, Phys. Rev. **88**, (1952) 184.

[19] N. Kolachevsky, P. Fendel, S. G. Karshenboim and T. W. Hänsch, *2S hyperfine structure of atomic deuterium*, Phys. Rev. A **70**, (2004) 062503.

dance with all theories of hydrogen atom, two atomic states with almost equal orbital radii ought to have energy levels (or spectral terms) whose hyperfine splitting can differ at most by a 2-factor.

Still this rule obligatory for hydrogen atom is never experimentally observed. The most evident example are the two energy levels $2S$ and $2P_{1/2}$, which differ from each other by only 0.00013%, and yet these two spectral terms extremely close in the Grotrian diagram exhibit hyperfine splits[14] in a ratio inexplicably high for two atomic states $2s$ and $2p_{1/2}$ with almost equal radii, $\Delta v_{2S}/\Delta v_{2P_{1/2}} = 1.77556 \cdot 10^8$ Hz$/0.59184 \cdot 10^8$ Hz $= 3$.

(4) The fictitious orbital dipole magnetic field fabricated by Pauli, Goudsmit and Uhlenbeck in the hydrogen atom has a similarly fictitious orbital magnetic moment $M_{orb} = I_n S_n$, where I is the intensity of the electric current formally represented by the orbital motion of the atomic electron, S is the area delimited by the atomic orbit, and $n = 1, 2, 3 ...$ is the main quantum number introduced by Bohr for the energy levels in the hydrogen atom. Or, as $I = ev_n/2\pi r_n$, $S = \pi r_n^2$ and $mv_n r_n = n\hbar$, it results an orbital magnetic moment $M_{orb} = ev_n r_n/2$, equal to $M_B = eh/4\pi m_0$ in fundamental atomic state $1s$ with quantum number $n = 1$, to $2M_B$ in $2s$ atomic state with $n = 2$, to $3M_B$ in $3s$ atomic state, etc., so that in their highly excited states the Bohr's atom would reach magnetic moments of hundreds Bohr magnetons! And because these imaginary orbital dipole magnetic fields with axial symmetry have their symmetry axes almost coincident with the symmetry axis of the dipole magnetic field generated by the central proton permanently placed very close to the geometric center of the atomic orbits, in the hydrogen line spectrum the fine splitting of any P, D or F spectral term and the hyperfine splitting of all its fine subterms have to be in a ratio approximately equal to that of the magnetic moments had in each atomic state by the two intra-atomic fields responsible for these spectral splits.

However, the fine splitting $\Delta v_{2P} = 10{,}969.1$ MHz of the spectral term $2P$ and the hyperfine splitting $\Delta v_{2P_{1/2}} = 59.19$ MHz of its fine subterm $2P_{1/2}$ are in a ratio $\Delta v_{2P}/\Delta v_{2P_{1/2}} = 185$ much smaller than the corresponding tio $2M_B/M_p = 1314$ between the orbital magnetic moment of atomic electron $M_{orb} = 2M_B = 2 \times 9.273 \cdot 10^{-24}$ A$\cdot$m^2 allegedly responsible for the fine splitting Δv_{2P} and proton magnetic moment $M_p = 1.420 \cdot 10^{-26}$ A$\cdot$m^2 assuredly responsible for the hyperfine splitting $\Delta v_{2P_{1/2}}$, although the two atomic states $2p$ and $2p_{1/2}$ have practically identical energies and orbital radii, and this discrepancy is much too large to be explained by the possibly different geometrical factors $\sqrt{(1+3\cos^2\theta)}$ of the two p atomic states.

(5) Since the line spectra assigned to hydrogen and deuterium atoms are almost identical, which proves an almost identical structure of these atoms in all their homologous atomic states, the hyperfine splitting of their homologous spectral terms ought to be in a ratio equal to that of the two known magnetic moments of their nuclei, proton and deuteron, $M_p/M_d = 1.410 \cdot 10^{-25}$ A·m$^2/0.433 \cdot 10^{-25}$ A·m$^2 = 3.252$. But again inexplicably, although the homologous fundamental terms $1S$ and $2S$ in the two line spectra assigned to the atoms of light and heavy hydrogen have hyperfine splits in equal ratios, $\Delta v_{1S(H)}/\Delta v_{1S(D)} = 1.420406 \cdot 10^9$ Hz$/3.27384 \cdot 10^8$ Hz $= 4.339$ and $\Delta v_{2S(H)}/\Delta v_{2S(D)} = 1.77556 \cdot 10^8$ Hz$/4.09245 \cdot 10^7$ Hz $= 4.339$, as normal, this ratio is significantly different from that expected 3.252, and this incomprehensible difference is totally incompatible with the two almost identical line spectra assigned to hydrogen and deuterium atoms.

Finally two clear conclusions result from these generalized discrepancies between experimental reality and what all atomic theories predict regarding the fine and hyperfine splitting of energy levels in hydrogen line spectrum:

(1) The fine splitting of the energy levels of the radiating electron in hydrogen atom is impossible, because such an effect could appear only in the dipole magnetic field generated by another electron, evidently non-existent in hydrogen atom. This very simple but indisputable truth proved right from the very outset the molecular origin of the line spectrum assigned by Bohr to atomic hydrogen without the slightest evidence, because the second electron able to cause the fine splitting of the energy levels of the radiating electron exists only in hydrogen molecule. It is beyond doubt, the orbital magnetic moment invented in despair of cause has been just a rudimentary manipulation, the orbital motion of atomic electrons cannot generate dipole magnetic fields defined by magnetic moments;

(2) All experimentally measured fine and hyperfine splits of the energy levels assigned to hydrogen atom confirm the above conclusion, because all of their values are in ratios inexplicable for a structure so simple as that of the hydrogen atom made of only one central proton and only one electron in orbital motion around the nucleus-proton. As it has been shown, if the spec-

tral terms in the hydrogen line spectrum would belong to atomic hydrogen, all their magnetic splits ought to be directly proportional with magnetic induction experienced by their radiating electrons in the different atomic states, which can be calculated from the known magnetic moment of the field causing one or other spectral splitting and the orbital radius predicted by atomic theories for each atomic state. If these obligatory correlations between the experimentally measured spectral splits and those predicted on the basis of atomic theories do not exist, the line spectrum of hydrogen passed through discharge tubes cannot belong to the free H atoms.

In both cases the only alternative is the molecular origin of the line spectrum always assigned to atomic hydrogen, another explanation simply cannot exist. In other words, gaseous hydrogen passed through a discharge tube has to be composed of excited but not dissociated H_2 molecules.

Doubtlessly such a groundbreaking conclusion makes inevitably necessary a critical review of all arguments claimed even today in favor of this primary belief in the dissociated state of gaseous hydrogen supplied by a discharge tube. Fortunately this is not too hard, because these arguments are entirely unsubstantial.

Strange case of orthohydrogen magnetic moment

Atomic state of hydrogen subjected at very low pressure to electric discharges was presumed since the second half of the 19th century, but besides the Stern-Gerlach experiments with hydrogen its experimental ground is even today much too inconclusive in relation to its importance.

Actually only the Langmuir's torch and a higher chemical reactivity of hydrogen previously subjected to electric discharges can be found in specialized literature[20] as arguments in this respect. Still the Langmuir's torch has not a monochromatic photon emission, an obligatory characteristic for an exothermic recombination $2H \rightarrow H_2$, but a continuous spectrum mainly in

---

[20] Molecular hydrogen has a vast literature, here only some comprehensive books:
P. Pascal, *Nouveau traité de chimie minérale*, Vol. I, Maison et Cie, Paris, 1956.
E.A. Moelwyn-Hughes, *Physical Chemistry*, Pergamon Press, Oxford, 1961.
K. H. Mackay, *The Element Hydrogen, Ortho- and Para-Hydrogen, Atomic Hydrogen, in Comprehensive Inorganic Chemistry*, Pergamon Press, Oxford, 1975.
R. B. Heslop and K. Jones, *Inorganic Chemistry*, Elsevier, Amsterdam, 1976.

the ultraviolet domain, therefore the huge heat evolved at the place of contact between the hydrogen jet and the surface of refractory metals is rather emitted by de-exciting in successive stages the H_2 molecules previously excited by electric discharges. Similarly, the higher chemical reactivity of gaseous hydrogen passed through a discharge tube can also be explained by its high content in excited molecules, more reactive because of their higher energy.

Moreover, a monochromatic radiation with energy equal to the binding energy of the H_2 molecule has never been noticed, neither at Langmuir's torch, nor in a Stern-Gerlach apparatus or in hydrogen maser, although in all these cases the H atoms assumed to occur in the discharge tubes go finally into common atmosphere, where their lifetime is very short and the molecular state of hydrogen is the only possible. Well, this generalized absence of such a monochromatic radiation has been justified by a theory according to which two free hydrogen atoms cannot recombine by their simple collisions because the energy evolved by their binding is equal or a little higher than the energy necessary for dissociating again the new formed H_2 molecule. In other words, as always such a recombination energy is evolved under the form of energy quanta $\varepsilon = h\nu$, it follows that the new formed H_2 molecules absorb instantaneously these quanta of energy just emitted by them and consequently dissociate again, so that the number of the free H atoms does not decrease on aggregate. More explicitly, these energy quanta are emitted by atomic electrons during their pairing in a covalent link, but the same electrons re-absorb them immediately after emission; we have here, therefore, some magic electrons able to absorb even the photons emitted by themselves! However, conforming to this theory two H atoms in collision can recombine yet, but only through three-body collisions $H + H + M$, when the energy evolved by their recombination is taken over by a third body M, which converts it into its kinetic energy.

And thus it has been explained not only why the energy quanta emitted during $H + H \rightarrow H_2$ re-combinations are never noticed, although their emission in such a process cannot be doubted, but also the unexpectedly high lifetime of the presumed free H atoms, experimentally found to reach even 0.5 s in certain circumstances, anyway extremely higher than their lifetime predicted by the kinetic theory of gases: these so long lifetimes of the free atoms would be just a consequence of the fact that the three-body collisions $H + H + M$, postulated as being the only effective for definitive re-combinations $2H \to H_2$, have a probablity of about 10^3 times smaller than the simple collisions $H + H$. But how this so long lifetime of the presumed free H atoms has been determined? Very simple, by measuring the extinction time of the noticed Balmer spectral lines, whose emission by atomic hydrogen was previously postulated by the Bohr's atomic theory!

Accordingly, first time the whole line spectrum of hydrogen is asigned be decree to atomic hydrogen, and then we confirm the atomic state of gaseous hydrogen coming out from the discharge tubes just by noticing the Balmer lines in its spectrum! What a perfect circular demonstration!

Differently, if gaseous hydrogen coming out from discharge tubes contains only excited H_2 molecules, and not free H atoms, as the fine and hyperfine splitting of the hydrogen spectral terms proves as clear as possible, everything becomes much simpler: no ultraviolet monochromatic radiation expected from $2H \to H_2$ re-combinations has been noticed simply because no such a recombination takes place, and consequently the long extinction time noticed for the Balmer spectral lines shows in fact a trivial fluorescence in molecular hydrogen previously highly excited through electric discharges at very low pressure, noticed also to many other substances!

Therefore, the only argument regarding the dissociated state of hydrogen passed through a discharge tube seems to remain that magnetic moment equal to one Bohr magneton found by Stern and Gerlach in their experiments with hydrogen, which cannot belong to molecular hydrogen, as long as conforming to the Pauli's principle formulated in 1925 neither parahydrogen nor orthohydrogen can be paramagnetic.

Still this quantum principle seemed to be invalidated in the 1930's, after discovering the so-called *electronic spectrum* of molecular hydrogen, because this spectrum exhibits two distinct scales of discrete spectral terms, without that band structure proper in general for all molecular spectra (excepting a small continuous domain in one of the two discrete scales)[21].

Or, such a double scale is proper for all atoms, ions and molecules with two electrons on the last electronic layer (the He atom is the most salient case) and its origin is very well known: one of the two scales has singlet

---

[21] G. Herzberg, *Molecular Spectra and Molecular Structure*, Vol. I, Van Nostrand, Princeton, 1965.

M. A. Ellyashevich, *Atomic and Molecular Spectroscopy*, GIFML, Moscow, 1962. (in Russian)

terms and belongs to their diamagnetic paraisomers with antiparallel spins $S = 0$, while the second contains triplets and belongs to their orthoisomers with parallel electron spins $S = 1$, whose paramagnetism is thus explicit.

But because the reality of paramagnetic ortho-H_2 molecules having a ratio magnetic moment/mass identical to that of the paramagnetic H atoms, and consequently with the same trajectory in the Stern-Gerlach apparatus!, would have meant an unexpected but definitive denial of the Pauli's principle, fundamental for the new atomic and molecular quantum physics then in full swing, all spectra with two scales $S = 0$ and $S = 1$ were essentially distorted by ruling out the fundamental term of the scale $S = 1$, even if this nonsensical amputation entails a multiple violation of the rule $\Delta S = 0$ otherwise strictly observed, even if the ortho-H_2 molecules and the He atoms with only nuclear magnetism in their fundamental state become so suddenly paramagnetic in their excited states, and even despite the well known energy difference between the unexcited ortho- and para-H_2 molecules, a reality proved by all experimental data concerning their reversible para ↔ ortho conversions!

Obviously all these nonsensical oddities of this Grotrian diagram with two scales $S = 0$ and $S = 1$, but only one fundamental term $S = 0$, are the bitter price paid for preserving no matter how the nuclear magnetism of orthohydrogen, at least in its unexcited state (as tough an excited H_2 molecule $S = 1$ could remain not dissociated despite the strong repulsive interaction acting in accordance with quantum theory between its two electrons with parallel spins). But who could think of a huge confusion between paramagnetic orthohydrogen and paramagnetic atomic hydrogen after the great success of the Bohr's atomic theory?

Moreover, the postulated nuclear magnetism of orthohydrogen was never experimentally confirmed, although such a demonstration would have been entirely possible right from the start. Indeed, already in the 1930's Estermann, Frisch, Stern and others measured accurately enough the proton magnetic moment in a Stern-Gerlach type apparatus with adequately increased gradients of magnetic field, but failed in the case of orthohydrogen for reasons never elucidated. After all, why the proton magnetic moment could be measured in this way, but not that one of orthohydrogen molecule, although conforming to quantum theory both particles ought to have the same ratio magnetic moment/mass, and therefore the same trajectory not only inside the non-uniform magnetic fields in the Stern-Gerlach equipments, but also inside the uniform magnetic fields used in magnetic resonance experiments? Or, if for these unsuccessful attempts in the case of the Stern-Gerlach method we can think that the fault was the using of high magnetic gradients suitable only for particles presumed to have just nuclear magnetism, but absolutely inadequate in the case of paramagnetic orthohydrogen, the absence of an experimentally measured magnetic moment of orthohydrogen even after discovering the method of nuclear magnetic resonance becomes something entirely incomprehensible. Indeed, all magnetic moments of fundamental particles with nuclear magnetism (proton, neutron, deuteron, triton, muon, charged pion, etc.) have for long been very accurately measured by this method, but orthohydrogen has been forgotten inexplicably.

And this amnesia has not disappeared until now, so that today none of the numerous books or online sites dedicated to hydrogen mentions an experimentally measured magnetic moment of orthohydrogen in their comprehensive lists of physical properties of molecular hydrogen. Or, is there something that cannot be known by the physicists' community?

Let us resume the two antagonistic versions regarding the magnetism of molecular hydrogen:

(1) Molecular quantum physics postulates the diamagnetism of the para-H_2 molecule whose electrons have antiparallel spins and magnetic moments, but the nuclear magnetism of the ortho-H_2 molecule because the two electrons in this molecule have also antiparallel spins, while its two protons have parallel spins and magnetic moments;

(2) The two distinct scales in the electronic spectrum of molecular hydrogen prove the diamagnetism of parahydrogen whose spectral terms are exclusively singlets, because the para-H_2 molecule has both electrons and protons with antiparallel spins and magnetic moments, but the paramagnetism of orthohydrogen whose spectral terms are exclusively triplets, because the ortho-H_2 molecule has both electrons and protons with parallel spins and magnetic moments.

It is very clear, as long as even now orthohydrogen has not a clearly determined magnetism, in spite of the available modern methods able to measure very accurately its magnetic moment, foremost that of magnetic resonance, the dissociated state of hydrogen passed through discharge tubes cannot be attested by the Stern-Gerlach experiments with hydrogen, and consequently the molecular origin of the hydrogen line spectrum assigned by Bohr to atomic hydrogen remains a variant with no sound experimental counterargument.

Microwaves $\lambda = 0.211$ m from hydrogen maser and cosmic space

If the whole line spectrum assigned to the free H atoms is emitted in fact by molecular hydrogen, then the generalized hyperfine splitting of all its S spectral terms and P, D and F fine subterms becomes a consequence of the very small energy difference between the ortho- and para-H_2 molecules. This means evidently a new origin of the famous $\lambda = 0.211$ m microwaves, which become so a monochromatic radiation emitted through a hyperfine magnetic interaction electron-proton during the slightly exothermic ortho → para conversion of molecular hydrogen.

And indeed, numerous experimental researches in the field offer sound arguments for such a version. Thus, in compliance with these experimental studies we can think that in hydrogen maser the incident normal hydrogen containing about 75 % orthoisomer is initially much enriched in this isomer of higher energy by passing it successively, exactly as in the Stern-Gerlach experiments with hydrogen, first time through a discharge tube where the high temperature increases its content in orthoisomer, then through a strong non-uniform magnetic field having a similar effect, and finally its return to the normal composition inside the central quartz bulb is much accelerated by coating the inner walls of this highly vacuumed receptacle with superficially active substances, as Teflon or paraffin, able to adsorb temporarily the ortho-H_2 molecules in excess and to cause their exothermic conversion into the para form during these short-lived adsorptions. And because the energy difference between the ortho- and paraisomers of molecular hydrogen is very small, this monochromatic radiation emitted during ortho → para conversions has surely to be situated in the range of microwaves with long wavelengths. Also, if hydrogen maser radiation $\lambda = 0.211$ m is emitted by this molecular mechanism, its Doppler broadening has to be null, because the emitting ortho-H_2 molecules are at rest on the inner walls of the central bulb. Differently, since the cosmic dust, whose very small and rough granules rich in paramagnetic atoms can also adsorb ortho-H_2 molecules and induce their conversion into the para form, is always in motion, universal radiation $\lambda = 0.211$ m emitted in cosmic spaces has to have a variable Doppler broadening. This ability of the molecular version to explain naturally such an essential difference between hydrogen maser radiation with no spectral width and cosmic radiation $\lambda = 0.211$ m with Doppler spectral width is an important argument in its favor. In addition, some questions without explanations until now become easy to understand when we have in view some specific features of the inverse conversion para- $\rightarrow$ orthohydrogen, which is evidently caused by absorptions of the microwaves $\lambda = 0.211$ m existing in the surrounding space by the para-H_2 molecules:

(1) The absence of a universal radiation emitted by cosmic deuterium, because, differently from the case of light hydrogen, in spite of its higher energy orthodeuterium is the stable isomer at low temperatures, and so a natural exothermic ortho $\rightarrow$ para conversion of deuterium has a very small probability in cosmic space;

(2) Why pure parahydrogen kept in closed vessels at very low temperature experiences yet a very slow but discernible conversion into ortho form, even if its temperature is maintained rigorously constant;

(3) Why pure parahydrogen can be kept long time in the closed vessels in laboratory, but changes quickly into normal hydrogen when reaches the open space outside laboratory.

Obviously, in the two last cases both the slow conversion para $\rightarrow$ ortho in closed containers maintained rigorously at the lowest possible temperature and that much faster in open spaces is caused by universal radiation $\lambda = 0.211$ m existing everywhere in Universe, whose density in closed spaces is evidently much lower than within the free spaces, but never absolutely null.

On the other hand, quantum theory of microwaves cannot reliably explain why hydrogen maser radiation $\lambda = 0.211$ m has no spectral width, neither Doppler nor natural:

(1) Any monochromatic photon radiation emitted by free hydrogen atoms in motion with velocity about $2.5 \cdot 10^3$ m/s has inevitably to exhibit a Doppler broadening of about 10^4 Hz, this is an elementary aspect always experimentally noticed;

(2) Conforming to the Heisenberg's equation of uncertainty, the null natural broadening of hydrogen maser radiation requires a really huge lifetime $\Delta t \geq 1$ s of the excited state responsible for its emission, therefore of about 10^9 times higher than all lifetimes measured by delayed coincidence for excited atomic states of hydrogen.

But, as in many other cases, when experimental reality proved to be in disagreement with quantum theory, all inconvenient facts have been covered by ad hoc explanations, sometimes of a crazy fun. For example[22], hydrogen maser radiation $\lambda = 0.211$ m has no Doppler broadening simply because within the quartz bulb the free hydrogen atoms change very quickly their motion direction by reflection on the inner walls, while the lack of any natural spectral width of hydrogen maser radiation is explained by equalizing the lifetime of the excited state responsible for its emission with the staying time of the free hydrogen atoms inside the quartz bulb, arbitrarily considered to be about 1 s !

And conforming to another fanny version invented for justifying why hydrogen maser radiation emitted by atoms in motions has no spectral width, this absence is a consequence of the fact that the quartz bulbs used in hydrogen maser have diameters smaller than the wavelength $\lambda = 0.211$ m of the emitted microwaves. In other words, when such a bulb has a diameter, say, equal to 0.22 m , hydrogen maser radiation can have spectral width, but not when its diameter is only 0.20 m !

Evidently, all these figments ignore with serenity all similar situations where the Doppler broadening of spectral lines always appears. And the best example in this sense is just the common line spectrum of hydrogen, also emitted by hydrogen previously passed through a discharge tube just in quartz receptacles at room temperature and very low pressure.

It is very clear, only the special coat of the quartz vessel in hydrogen maser, which is the only difference from those where the common spectrum of hydrogen is recorded, can really cause the null Doppler broadening of the $\lambda = 0.211$ m microwaves emitted here, and the mechanism can only be a temporary adsorption of the emitting particles on the special inner surfaces of the quartz bulbs used in hydrogen maser, during of which these emitting particles at rest emit naturally microwaves with no Doppler width.

---

[22] G.-C. W i c k, *The Extention of Measurement*, Suppl. Nuovo Cim. **4**, (1966) 309.

Differently, quantum theory of hydrogen maser radiation cannot take into account such a logical emission without Doppler broadening during a short adsorption of the supposed hydrogen atoms on the Teflon or paraffin inner coating of the quartz bulb, simply because such an adsorption of the free atoms on the inner walls of the quartz bulbs can last at most 10^{-5} s at room temperature, and in turn this maximum adsorption time would mean a similar lifetime of the excited state considered in the Heisenberg's relation of uncertainty, therefore a minimum natural width of about 10^5 Hz for hydrogen maser radiation $\lambda = 0.211$ m !

Besides all these manifest contradictions, quantum theory of hydrogen maser radiation involves right from the start another one inexplicably overlooked until now. If hydrogen atoms are of two kinds, one made of a proton and an electron with antiparallel spins, therefore with a higher energy than the other kind, made of a proton and an electron with parallel spins, and all hydrogen molecules have their electrons paired exclusively with antiparallel spins, then molecular hydrogen ought to have three isomeric forms:

(1) H_2 molecules made of two H atoms of smaller energy;

(2) H_2 molecules made of two H atoms of higher energy;

(3) H_2 molecules made of two different H atoms.

Evidently the two first structures would have antiparallel nuclear spins, and therefore they would be two distinct species of parahydrogen, while the third would have parallel nuclear spins, and therefore would correspond to orthohydrogen. More, the energy of orthohydrogen would be intermediary between those of the two forms of parahydrogen!

Obviously, just two isomers of molecular hydrogen can be formed only through two different binding manners of two identical H atoms, and consequently in the diamagnetic para-H_2 molecules the two valence electrons have certainly an antiparallel mutual orientation of their spins and magnetic moments, while in the ortho-H_2 molecules the two electrons have inevitably parallel spins and magnetic moments. Another explanation for the only two isomeric forms of molecular hydrogen simply cannot exist.

Another very important but still unclear question is the energy difference between the two isomers of molecular hydrogen, because the energy released during exothermic conversion ortho- → parahydrogen has never been measured, and nor yet estimated from continuous spectrum of molecular hydrogen, where the two isomers have different peaks. For this reason, we find in literature for this conversion energy only values calculated by the methods of physical chemistry, whose magnitude order is out of an elementary reality: whatever mechanism for ortho → para conversion is taken into account, with or without dissociating the initial ortho-H_2 molecule, the energy evolved during its ortho → para conversion cannot exceed the energy proper to one hyperfine magnetic interaction within hydrogen atom, just because in quantum atomic and molecular physics the ortho → para conversion of a H_2 molecule consists only in an inversion of a proton spin within one of its H atoms. Indeed, the conversion energy calculated by physical chemistry varies between about 500 kJ/kg at very low temperatures and about 30 kJ/kg at normal temperatures, which utterly disagree with the energy about 0.57 kJ/kg of the $\lambda = 0.211$ m microwaves issued in hydrogen maser, and this huge disproportion is absolutely incomprehensible:

(1) If the ortho → para conversion of a H_2 molecule involves a spin inversion in one of the two free hydrogen atoms resulted by its temporary dissociation on an active surface, then its energetic effect has evidently to be an emission of a microwave $\lambda = 0.211$ m ;

(2) If this ortho → para conversion involves a nuclear spin inversion inside an undissociated H_2 molecule, the released energy would be even much smaller, because according to quantum theory in all H_2 molecules the two electrons have on aggregate a null magnetic field because of their pairing with antiparallel spins and magnetic moments, so that any flip-flop of a proton magnetic moment within a H_2 molecule takes place only inside the weak magnetic field of the other proton, and such an *ultrahyperfine* magnetic interaction proton-proton can have only an extremely small energetic effect. Also, how could the external temperature change so much the energy released by a spin reversal within the atom?

Anyway, this lack of interest for determining the energy variation in the reversible conversions ortho ↔ para of H_2 molecules is difficult to understand, the more so as this very important but still unknown physical quantity could be obtained easy enough by measuring the frequency of monochromatic radiation emitted when orthohydrogen of high purity (which can be obtained by special methods) passes into para form, for example during its adsorption in a vacuumed space on a surface doped with paramagnetic atoms, by using an ordinary meter able to detect microwaves in the frequency range 1 ... 3 GHz .

On the other hand, the molecular origin of the $\lambda = 0.211$ m microwaves raises a clue question: as these microwaves result indubitably from hyperfine magnetic interactions electron-proton, why do the fine magnetic interactions electron-electron not appear yet when the paramagnetic ortho-H_2 molecules change into the diamagnetic para-H_2 molecules, although such a conversion involves evidently a change of the mutual orientation of the angular and magnetic moments of the two valence electrons, from parallel into antiparallel?

Well, the laws of electromagnetism can specifically explain the absence in well-defined conditions of the fine magnetic interactions electron-electron when paramagnetic ortho-H_2 molecules (whose electrons should have parallel magnetic moments) change into diamagnetic para-H_2 molecules (whose electrons have antiparallel magnetic moments). Indeed, if quantum theory of covalent link postulates very simplistically "A pair of valence electrons with antiparallel spins ($S = 0$) denotes *attraction*, while a pair with parallel spins ($S = 1$) denotes *repulsion*."[23], conforming to the basic laws of electromagnetism, two electrons have always an attractive magnetic interaction when in the space between them the flux lines of their dipole magnetic fields are convergent, even if their magnetic moments are not antiparallel. More explicitly, a covalent link between two electrons can be formed either by two very close electrons with collinear and parallel magnetic moments, or as well by two very close electrons with magnetic moments antiparallel, but both perpendicular to the common symmetry plane Oxy of their two dipole magnetic fields. First configuration could be proper to paramagnetic orthohydrogen, the second to diamagnetic parahydrogen.

More, if in the ortho-H_2 molecules the distance between the two magnetically paired electrons is $\sqrt[3]{2} = 1.260$ times larger than in the para-H_2 molecules, then in both isomeric forms each electron would experience the same magnetic induction generated by the second electron in molecule, and consequently all the ortho $\leftrightarrow$ para conversions of the H_2 molecules take place with no fine interaction between the two molecular electrons, but with two hyperfine interactions between each electron in one H atom and the proton in the other atom. Evidently this ratio $\sqrt[3]{2}$ comes from the known equation $B = \sqrt{(1 + 3\cos^2\theta)}\,\mu_0 M / 4\pi r^3$ of the dipole magnetic magnetic fields

---

[23] V. N. Kondratiev, *The Structure of Atoms and Molecules*, Mir, Moscow, 1967.

with axial symmetry generated by all elementary particles, including the electrons and protons, because it corresponds to a pair of points with equal magnetic induction B in the same dipole magnetic field, one point placed on the symmetry axis Oz ($\theta = 0$) of the considered dipole magnetic field, and the other point on any Ox or Oy axis ($\theta = \pi/2$) of that field.

Amazingly, the coaxial or parallel coupling of the dipole magnetic moments that belong to the two electrons in H_2 molecules was anticipated by Parson[24] since 1916, before discovering the two isomers of molecular hydrogen!

Too many grave discrepancies ignored for too long

Probably first reply of a physicist after reading all arguments presented in first chapter for halving numerical value of the Planck's constant would be like this: "Very well, but how can such a late revaluation be accepted as long as numerous branches of physics built up on the present value $h = 6.626 \cdot 10^{-34}$ J·s of this fundamental constant have been confirmed by an overwhelming basis of experimental data, as all treatises of physics claim unreservedly?". And almost certainly, atomic physics would be the first mentioned in this respect, whereas most of its specific equations contain, explicitly or implicitly, the action constant h.

Unfortunately, as many times behind the show case the things are not at all so bright, at a critical and unprejudiced reexamination atomic physics proves to be very far from its alleged condition of absolute milestone in the knowledge of microcosm, as long as its basic hypothesis, namely the atomic state of hydrogen subjected to electrical discharges, proves to be rather a chimera coming from the prehistory of microcosm physics, anyway in clear disaccord with the fine and hyperfine spectral splitting noticed in the hydrogen line spectrum. Experimental data are very clear in this respect, even if up to now it has been preferred to dissimulate their very inconvenient significance by all means, for instance by inventing a fictitious dipole magnetic field generated by the orbital motion of atomic electron, able to split even the energy of its elementary source, while the clear incompatibility between the measured hyperfine splits of the spectral terms assigned to hydrogen

---

[24] A. L. Parson, *A Magneton Theory of the Structure of the Atom*, Smithsonian Miscellaneous Collection **65**, 11 (1915).

atom and the very simple structure of this atom has been deliberately ignored, as though it simply would not exist. And then many other deliberate distortions had to be accepted for saving quantum atomic physics, either in order to rule out the orthoisomer of molecular hydrogen as a valid alternative for atomic hydrogen in explaining the Stern-Gerlach experiments with hydrogen, or to justify why hydrogen maser radiation has no Doppler width despite its emission by moving atoms.

But in good measure all these grievous misrepresentations have resisted until now because of a very strange (euphemistically said!) lack of interest for measuring accurately, by methods available for a long time, two amazingly still unknown properties of orthohydrogen, essential for atomic and molecular physics, and in last instance for the whole quantum physics, namely its magnetic moment and the energy released during its exothermic conversion into para form, therefore exactly the ones able to confirm or infirm definitively the presumed dissociated state of hydrogen coming out from a discharge tube. No rational explanation can justify such a lack of interest, and inevitably this reality raises serious queries about what really happens in physics of the last decades.

Anyway, without a full will to establish definitively the atomic or molecular state of hydrogen supplied by discharges tubes, it is very clear that all atomic theories developed until now, including the Bohr's atomic theory used in the 1920's for eliminating the spinning ring electron, will remain under the justified suspicion of a terrible confusion between the free H atoms and the ortho-H_2 molecules.

ELEMENTARY EVIDENCE AGAINST SPECIAL RELATIVITY

Nuclear and atomic mass defects

The mass equation $m = m_0/\sqrt{1-v^2/c^2}$ was included in special relativity in 1905, when it already had a first experimental confirmation (Kaufmann, 1901), and in the next years many other experiments with β-electrons and electrons accelerated in electromagnetic field confirmed its validity for almost entire scale of velocities. However, the mass measurements in nuclear physics posed an unexpected and very annoying problem: all nuclei have rest mass below the summed rest masses of their constituent nucleons. As example, the rest mass of deuteron $m_d = 2.01355$ u is a little smaller than the sum of the rest masses $m_p = 1.00728$ u and $m_n = 1.00867$ u of its constituent proton and neutron, and this very small difference of mass $\Delta m = m_p + m_n - m_d = 0.00240$ u corresponds exactly to the binding energy of deuteron. Therefore, the mass of at least one of the two nucleons in deuteron is smaller than the rest mass of that nucleon, although conforming to relativistic interpretation of the equation $m = m_0/\sqrt{1-v^2/c^2}$, the mass m of an elementary particle can never be smaller than its rest mass m_0, simply because always $v < c$ and $1/\sqrt{1-v^2/c^2} > 1$.

As the very influent supporters of special relativity could not accept such a sudden fall of their unassailable theory, the relativists had to find an explanation for this nuclear mass defect absolutely incompatible with the relativistic meaning of the mass equation $m = m_0/\sqrt{1-v^2/c^2}$, and therefore able to invalidate the whole theory. The found solution was the liquid drop model of nuclei.

Conforming to this vision, differently from the atoms made of distinct electrons in orbital motion around a nucleus, nuclei cannot be discrete structures made of distinct nucleons in motion because of the high strength of the intra-nuclear interactions, which would change them into viscous quasi-uniform entities, wherein the component nucleons lose their characteristic functions had in the free state, including a determinate mass always higher than their rest mass. However, the inconsistency of this invocation of the strong intensity of the intra-nuclear forces acting between nuclear nucleons is easy to prove. For example, all K electrons ($n = 1$) inside the heaviest atoms ($Z > 100$) have binding energies about 10^5 times greater than those of the marginal electrons in these atoms, so that their binding energies are already at the level of nuclear binding energies, but yet all these K electrons preserve their individuality and specific properties exactly as the valence electrons of these atoms. The fact that any K electron can be extracted from atom by an incident photon having the necessary energy, exactly as in any trivial photoelectric effect, proves in full this viewpoint.

For all that, the fervent supporters of special relativity adopted so ardently the drop model of nucleus, that even in 1950 "only the word *structure* of nucleus would have been considered criminal by most physicists"[25], although meanwhile the experimental nuclear researches adduced more and more evidence in favor of the individual motions of the nuclear nucleons, some of them (for instance[26]) very soon after discovering the nuclear mass defect (Aston, 1927).

And in the next decades the proofs in this sense became impossible to be ignored, such as the determining role of the orbital motion of all constituent nucleons in establishing the nuclear spin and magnetic moment, or in lowering substantially the energy threshold in the antinucleon engendering on nuclei, or in the birth of mesons on nuclei:

"As it is evident from the previous considerations, the birth of the mesons on the nuclei can be interpreted as a result of the birth of the mesons on the individual nucleons in nuclei. Thus, the energy spectra of the mesons could be acquired starting from the known section of the meson birth on free nucleons, by considering the internal motion of the nuclear nucleons. (...)

The circumstance is a proof in favor of the conception of the meson birth in odd nucleon-nucleon collisions within the nucleus. We provide also other results which in turn are proofs in favor of the hypothesis that the mesons are born in the nuclei through nucleon-nucleon collisions."[27]

---

[25] D. I. Blokhintsev, V. S. Barashenkov and B. M. Barbashov, *The structure of nucleons*, Usp. Fiz. Nauk **2** (1959) 505. (in Russian)

[26] L. Szilard and T. A. Chalmers, *Detection of Neutrons Liberated from Beryllium by Gamma Rays: a New Technique for Inducing Radioactivity*, Nature, **134** (1934) 494.

[27] L. M. Barkov and B. A. Nikolski, П – *Mesons*, Usp. Fiz.Nauk **61**, (1957) 341. (in Russian)

Experimental data proving the composite structure proton-neutron of the deuteron have also been very numerous:

"At the collision of the deuteron to nuclei, forming the composite nucleus is not compulsory. The *stripping* reactions are more probable, in which only one of the particles in the deuteron is absorbed and remains in nucleus, the second being a reaction product. (...)

Asymmetrical distribution of electric charge in deuteron also leads to the possibility of the electric splitting of deuteron, during which the proton and the neutron are simultaneously released. (...)

Since the dimensions of the deuteron are large, then the second particle may not reach the range of nuclear forces. This way, the capture of one of the particles in the deuteron corresponds unconditionally to releasing the other particle. (...)

Because of the low binding energy of the deuteron, in its interaction with nuclei the diffraction splitting that occurs far from nucleus becomes possible. (...)

Deuteron is a nucleus with a spatial structure. The spatial dimensions of the deuteron are characterized by the mean distance between its constituent neutron and proton."[28]

The Fermi (or kinetic) energies of the two constituent of deuterons are also proved in elastic scattering on the later[29], or by experimental values of the effective range of the $n-p$ triplet that show a composite structure of deuteron, which therefore cannot be regarded as an elementary particle like the proton or neutron.[30]

Clearly invalidated by experimental research, the liquid drop model of nucleus was at last abandoned after 1960, but meanwhile Yukawa thought a new theory of the nuclear forces acting between the nuclear nucleons, allegedly able to bring also into accord the nuclear mass defect and relativistic equation of mass. Inspired from electromagnetic interactions in quantum electrodynamics, in 1934 Yukawa presumed that the nuclear fields within

---

[28] A. G. Sitenko, *Interactions of deuterons with nuclei*, Usp. Fiz. Nauk **2**, (1959) 195. (in Russian)

[29] E. Ferreira, L. P. Rosa and Z. D. Thomé, *Effects of Fermi motion on deuterium scattering near a resonance*, Nuovo Cim. A **20**, (1974) 277.

[30] S. Weinberg, *Evidence That The Deuteron Is Not an Elementary Particle*, Phys. Rev. B **137**, (1965) 672.

nuclei act on the very close nucleons through their still undiscovered quanta, charged or neutral, but always unstable, and called *mesons* because their mass should have had to be somewhere between 200 and 300 electronic masses m_e.

This Yukawa's mesonic theory of nuclear interactions could briefly be explained in the following way[31]:

"According to quantum mechanics, the uncertainty relation $\Delta E \cdot \Delta t \approx \hbar$ states the maximum energy variation ΔE of an isolated system in the time interval Δt ("violation of the energy conservation law" during the short time Δt). Thus, if the time interval Δt is small, then ΔE can be very large.

Presume that nuclear interaction is so strong (rapid) because it occurs in the time interval $\tau_{nucl} = \Delta t$. Then we can accept that in the short interval Δt, a virtual meson with mass $m = \Delta E/c^2 = \hbar/(\Delta t\, c^2)$ arises in the close proximity of the nucleon on account of the energy $\Delta E = \hbar/\Delta t$.

Unlike the usual particles which can freely travel in space and time, the virtual particles exist only in the short interval of time Δt, during they can reach at a distance a from nucleon, without passing beyond $a = c\Delta t$. After the time interval Δt, the virtual particle is "captured" again by nucleon. Thus, the nucleon will be considered to be surrounded by a cloud of virtual mesons, emitted and absorbed continuously. The radius of this mesonic cloud (mesonic "mantle") is equal to $a = c\hbar/\Delta E = \hbar/mc$.

The virtual meson can be absorbed not only by its "own" nucleon, but also by any other nucleon in the range of mesonic mantle. The mechanism of nuclear interaction consists in an exchange of virtual mesons from one nucleons to other.

Quantitative values for the time τ_{nucl} of nuclear interaction and the mass m of the virtual meson can easy be obtained if the size a is identified with the action radius of nuclear forces. Considering it equal to $2 \cdot 10^{-13}$ cm (more exactly, $1.4 \cdot 10^{-13}$ cm), Yukawa obtained $\tau_{nucl} = a/c = 0.7 \cdot 10^{-23}$ s , $\Delta E = \hbar/\Delta t = 1.5 \cdot 10^{-4}$ erg ≈ 100 MeV, and $m = \Delta E/c^2 \approx 200\, m_e$. So the meson presumed by Yukawa became the possible quantum of nuclear field."

Well, when the $\mu^{\pm}$-mesons (or muons) with mass about 203 m_e were discovered in 1938, they real existence was at once considered to be a bril-

---

[31] K.N. Mukhin, *Experimental Nuclear Physics*, Vol. II, Atomizdat, Moskva, 1974. (in Russian)

liant confirmation of Yukawa's hypothesis, still this initial elation disappeared quickly because experimentally the muons exhibited a very weak interaction with nuclei. But the Yukawa's theory came back in attention in 1947, when the $\pi^{\pm}$-mesons with mass about 273 m_e were discovered and acknowledged as the new real quanta of the nuclear field because of their strong interaction with the nuclear substance, and even quickly rewarded with the Nobel Prize in 1949.

However, how the nuclear mass defect appears in this virtual mechanism of nuclear interactions without infringing the relativistic mass equation? In short, the response would be the following.

All changes of virtual mesons have two stages: one when the so-called *heavy* and *light* mesons appear and the other when the nucleons come back to their normal states. During the first stage the law of energy and mass conservation is not observed, but conforming to Heisenberg's relations of uncertainty, for a very short time (of about 10^{-23} s) the nucleons can have masses both higher or smaller than their rest masses, as they absorb or emit virtual masses, and anyway, on the whole each change of mesons observes the laws of energy and mass conservation.

Actually, the imaginary virtual mesons solve nothing concerning the nuclear mass defect. Nuclear nucleons could have real masses alternatively higher and smaller than their rest masses only if the virtual mesons emitted and absorbed by them would have real masses, which is a contradiction by itself, because only the real mesons can have real masses. Or, real mesons with real mass appear only by violent collisions between nucleons of high energy, while the nuclear nucleons neither have so high energy, nor collide with each other. Also, the virtual masses of the imaginary mesons presumed to be changed between them by the nuclear nucleons cannot be added or subtracted to or from the real masses of the nucleons, otherwise all the other elements of the virtual mesons besides their virtual masses, as for example their virtual electric charges, should be also taken into account, and this would lead to evidently insurmountable difficulties, because a real neutron electrically charged or a neutral real proton cannot to be accepted yet, even if only for "a very short time".

Actually the virtual mesons imagined by Yukawa are unable not only to justify the nuclear mass defect, but even the attractive nature of forces induced by them between nuclear nucleons, which is their main task in nuclei. Indeed, how could a nucleon be driven to another by absorbing a virtual or real particle emitted towards it by that other nucleon? In other words, how can a momentum proper to a field particle change into another with opposite direction after its absorption by a nuclear nucleon? Obviously, if two nucleons exchange continuously corpuscles with mass and momentum, the consequence can only be their mutual distancing from each other, therefore a rejection effect, and not at all their attraction.

But perhaps the clearest invalidation of the Yukawa's virtual mesons is the absence of nuclei made exclusively of protons or neutrons. This reality contradicts an organic prediction of this theory, namely its equal applicability to any pair of nucleons, $(p-p) \equiv (n-n) \equiv (p-n)$, an essential and inseparable part in mesonic theory of nuclear forces, known as *the isotropic invariance of nuclear forces* or *the independence of the nuclear forces on the electric charges of the nuclear nucleons*. It is very clear, the absence of some purely protonic and neutronic nuclei still needs another theory of intranuclear interactions, able to explain why two or more identical nucleons cannot form a bound structure stable in time.

And the things stand even worse as regards the atomic mass defect, existent for exactly the same reasons as in the case of the nuclear mass defect. Just as the binding energy 2.22 MeV of deuteron is equal to that evolved when one proton and one neutron bind to each other, the binding energy 13.56 eV of hydrogen atom is equal to that evolved when one free electron and one free proton bind to each other, and in both cases the evolved energy is equivalent to the mass defect of the new formed bound system. Therefore, the atomic electrons can also to have masses smaller than their rest masses, exactly as the nuclear nucleons. And not at all surprisingly, this atomic mass defect has been justified again by inventing virtual particles, this time virtual photons, whose actions in this regard are similarly confused and incoherent as those of the virtual mesons in nuclei. However, if in nuclei the main task of the virtual mesons is to induce attractive forces between nucleons, and only subsidiary to justify the nuclear mass defect, in atoms the attractive forces between nuclei and electrons are assigned in all atomic theories to electrostatic interactions between positive and negative electric charges, so that the role of the virtual photons is here exclusively to save the relativistic equation of mass. Maybe for this reason the atomic mass defect has been rather overlooked in physics books, and anyway never claimed as a brilliant confirmation of special relativity, as it happens in the case of the nuclear mass defect.

Anyhow, after a time these hardly sustainable explanations of the nuclear and atomic defects through imaginary particles acting beyond the Heisenberg's postulated borders of the physical certainty were gradually abandoned, so that nowadays the nuclear mass defects experimentally measured are mostly justified by the absolute interchangeability between mass and energy, expressed by formulations like "special relativity shows that the energy is an additional source of mass", or, in other words, if the equivalent mass of the total binding energy of a nucleus is added to its mass, then the mass defect of that nucleus formally disappears. Of course, such ingenuous ideas forget innocently the total energy of the photons emitted in the surrounding space at every forming stage of that nucleus, exactly equal to the total binding energy now desired to be added to its mass for dissimulating its mass deficit.

Beyond all attempts to hide the truth, both the nuclear and atomic mass defects invalidate definitively the relativistic deduction and understanding of the mass equation $m = m_0/\sqrt{1 - v^2/c^2}$, and through this the whole special relativity.

Electron with variable rest mass and energy

In 1895 Zeeman discovered that in an external magnetic field B each spectral line of hydrogen becomes a pair of two very near lines separated by a frequency gap $\Delta v = 2 M_e B/h$, equivalent to energy gap $\Delta E = 2M_e B$. This formula was reinforced some few decades later, when the fine and hyperfine splitting of hydrogen spectral terms were discovered. Also, in the more recent experiment with "geonium atom"[32] all the free electrons in uniformly circular motion in a plane perpendicular to an uniform magnetic field B show the same constant magnetic splitting $\Delta E = 2M_e B$ of their energy E , regardless their linear velocity. Or, just because this magnetic splitting $\Delta E = 2M_e B$ of the energy of both atomic and free electrons depends only on the external magnetic induction B , either intra-atomic or macroscopic, but not on their velocities, it has to be admitted even for an hypothetical electron with null velocity:

---

[32] P. Ekström and D. Wineland, *The Isolated Electron*, Sc. Amer. **243** (1), (1980) 91.

"To understand the behavior of the electron in the geonium atom it is helpful first to consider a simpler, idealized system: an electron at rest in an uniform magnetic field. Such a field can be represented by flux lines that are parallel and evenly spaced. The energy of this stationary electron depends on the orientation of its magnetic moment with respect to the external field. The energy is minimum when the moment and the field are parallel, and maximum when they are antiparallel."[32]

If so, we have also to accept that the rest energy and the rest mass of the electron cannot be intrinsic, immutable properties of this particle, as quantum relativistic physics claims, per contra, they evidently have an infinite number of values, two for each value of the external induction $\bar{B}$, depending on the parallel $\bar{M}_e \uparrow\uparrow \bar{B}$ or antiparralel $\bar{M}_e \uparrow\downarrow \bar{B}$ orientation of the electron magnetic moment $\bar{M}_e$ as against $\bar{B}$, the only two possible orientations of the electron magnetic moment in an external magnetic field. Therefore, the relativistic equations of the rest mass and energy of the electron should be $m_{0(B)} = m_0 \pm M_e B/c^2$ and $E_{0(B)} = E_0 \pm M_e B$, where m_0 remains only the rest mass of a notional electron in an imaginary space destitute of any magnetic field $B = 0$, and $E_0 = m_0 c^2$ the relativistic rest mass of such a notional electron. And when a relatively slow electron initially in a space with negligible magnetic induction $B \approx 0$ enter a uniform magnetic field B, its mass $m \approx m_0$ becomes $m_0 \pm M_e B/c^2$ by absorbing or emitting one photon with energy $h\nu = M_e B$. Magnetic resonance is an already discussed example in this regard.

Therefore, even if some physicists use different euphemisms for this shift $\Delta E_0 = E_{0(B)} - E_0 = \pm M_e B$ of the rest energy of the electron after entering a uniform magnetic field B, say, the *coupling* energy particle-field (but then we should likewise assume its correspondent *coupling* mass $\Delta m_0 = m_{0(B)} - m_0 = \pm M_e B/c^2$!) , what is really important cannot be eluded: the electrons have variable rest masses $m_{0(B)}$ and energies $E_{0(B)}$, because both these physical properties of them depend on the local magnetic induction B, which can be absolutely null nowhere in the real space.

More than evident, this infinite number of rest masses and energies of only one electron is quite incompatible with the pointlike electron of quantum and relativistic physics, whose unique rest mass and energy are immutable, intrinsic properties of it. On the contrary, such a variable rest mass and energy is explainable only if the electron has an inner structure responsive to the external magnetic induction, as evidently the inner structure of the spinning ring electron is.

Free electrons accelerated in electromagnetic field

In the 19th century Maxwell imagined a theory of electromagnetic radiation whose primary concept is the electromagnetic wave, always transverse because the two vectors attached to this wave, one of electric field and the other of that magnetic, have a constant sinusoidal oscillation on directions perpendicular not only to each other, but also to the propagating direction of the wave. But unfortunately for this theory, now it is assuredly known the granular structure of electromagnetic radiation, whose field particles called photons have mass, energy, momentum and spin, but not own electric and magnetic fields. More exactly, an electromagnetic interaction between two elementary particles, for example two electrons, means first a photon emission of one electron, and then an absorption of the emitted photon by the other electron. And the fundamental laws of energy, momentum and angular momentum have evidently to be observed in each of the two distinct stages of this electromagnetic interaction at a distance.

Also, the transverse polarization of electromagnetic radiation is noticed just in certain cases, a typical example being the radio waves emitted through mechanisms of antenna by slow conduction electrons, but when the emitting electrons have higher and higher velocities, their electromagnetic radiation has a proportionally higher degree of longitudinal polarization, one of the first noticed cases in this regard being Tscherenkow radiation, whose almost 100 % longitudinal polarization was much discussed at its time. Still more, electromagnetic radiation emitted even by the same electrons can have extremely different polarizations depending on their emission conditions, from almost 100 % longitudinal polarization to accurately 100 % transverse polarization, the most telling example in this regard being synchrotron radiation.

For all that, although all the other physical theories coming from the prehistory of physics have been forgotten for long owing to their total disagreement with experimental data appeared in the modern age, the Maxwell's understanding of electromagnetic radiation is often used even in many relatively recent physics books for explaining the very important processes of electromagnetic acceleration by which all researchers obtain the high-energy elementary particles necessary to get into the depth of microcosms. And the question is why, only because of that so much praised "mathematical beauty" of the Maxwell's equations?

Certainly not, this singular perenniality of the very old and meanwhile experimentally disproved Maxwell's theory has a much prosaic cause: the total incapacity of special relativity to explain how the fundamental laws of momentum and angular momentum conservation are respected in all processes of photon emission and absorption. And one of the most significant peculiar case is the absorptions of the photons, the field particles of electromagnetic interaction, by the free electrons.

Well, special relativity no more no less interdicts such photon absorptions simply because in such so simple processes the relativistic equations of energy and momentum are incompatible with each other. And the simplest case is that of an electron initially at relative rest and then accelerated to a certain velocity v by absorbing a photon with frequency ν, when the two relativistic equations of energy and momentum conservation in this process are evidently incongruous:

$$h\nu = m_0 c^2 \left(1/\sqrt{1 - v^2/c^2} - 1\right)$$

$$h\nu/c = m_0 v/\sqrt{1 - v^2/c^2} \ .$$

But then, how special relativity explains the photoelectric effect involving the quasi-free electrons in metal lattices? The relativists' answer is truly amazing:

"The free electrons cannot absorb a light quantum, because the relativistic laws of energy and momentum cannot be simultaneously satisfied ...

Such an absorption is possible only in the presence of a "third" body capable of ensuring the simultaneous fulfillment of both conservation laws. In other words, the photon absorption is possible only by the "bound" electron in atom, or another system, for example a crystal."[33]

Therefore, according to special relativity a quasi-free electron in the atomic lattice of a metal can absorb a photon whose energy $h\nu$ is integrally converted into kinetik energy, but before leaving the metal lattice the new formed photoelectron can transfer a part of momentum $h\nu/c$ received from

---

[33] S. V. Vonsovski, A. V. Sokolov and A. Z. Veksler, *Photoelectric effect in metals*, Usp.Fiz.Nauk **56**, (1955) 477. (in Russian)

the absorbed photon to the whole atomic lattice, namely as needed for observing the relativistic laws of energy and momentum simultaneous conservation. Of course, this relativistic interpretation of photoelectric effect does not explain how an electron can get rid of a part of its momentum with no changes of its mass, velocity and energy, as long as special relativity cannot admit an own photon emission of the emitted photoelectron.

It is really hard to find an ineptitude bigger than this miraculous capacity of the photoelectron to take the whole energy $h\nu$ of an absorbed photon, but of only a small fraction of the photon momentum $h\nu/c$, as these inseparable two physical properties of the photon, energy and momentum, both of them assessed by the same photon frequency ν, could be cutted separately, as desired.

Anyhow this ridiculous relativistic interpretation of photoelectric effect deserves no attention, because all the free electrons electromagnetically accelerated in electronic guns, synchrotrons or in laser-beams disprove indubitably the alleged incapacity of the electrons to absorb photons if they have not near entities able to take, nobody knows how, a part of the absorbed momentum $h\nu/c$. For example, the fast electrons delivered by electronic guns are initially in free state at rest on the cathode surface, where they intercept the accelerating photons emitted by the conduction electrons of the oscillating currents inside the cathode. Also, in the electromagnetic cavities in synchrotrons the fast electrons in motion in a straight line are further accelerated by absorbing in flight the photons emitted by high-frequency oscillating currents in directions parallel to those of the electrons. In fine, the free electrons accelerated in the focus of laser-beams are once more a flat denial of special relativity.

Very curiously, although in specialized literature there are plenty of works about the electromagnetic acceleration of the free electrons in electronic guns, synchrotron cavities or laser-beams, all authors forget that special relativity prohibits strictly such a process. Can anyone guess why?

No indisputable proof for invariant velocity of light

In the course of time, the second postulate of special relativity, which presumes an invariant velocity of light in all inertial frames of reference, was tested in many experiments whose results have been presented in specialized literature almost exclusively in their relativistic interpretation, without bringing forward in parallel their classical solutions, for enabling thus an immediate comparison between them. That is why, such a comparative review of some of the most representative experiments claimed along the time as evidence for the second relativistic postulate can be very instructive.

(1) First experiments worthy to be mentioned in this regard are undoubtedly those performed by Michelson and Morley, present many decades in all books in the field, but almost forgotten nowadays. The reason of this silent abandonment is the very late revelation of a very simple classical solution to these experiments, based on the experimental observation that in the reference frame of the laboratory the light does not change its initial velocity c after its reflection on a motionless mirror. Indeed, if in this system the velocity of light is independent of propagating direction, in the Sun's referential (or that *at absolute rest*) it becomes $\bar{c} + \bar{v}$, where $\bar{v}$ is the Earth's velocity. Therefore, in the reference frame at rest the speed of light in the arm parallel to $\bar{v}$ is $c(1 + v/c)$ on going and $c(1 - v/c)$ on return, and the path covering times on going (Δt_1) and on return (Δt_2) are equal,

$$\Delta t_1 = l(1 + v/c)/c(1 + v/c) = l/c$$
$$\Delta t_2 = l(1 - v/c)/c(1 - v/c) = l/c,$$

identical with the covering l/c time in the reference frame of laboratory. And in the arm perpendicular to $\bar{v}$, the path covering times in the laboratory referential are also the same,

$$\Delta t_1 = \Delta t_2 = l\sqrt{(1 + v^2/c^2)}/c\sqrt{(1 + v^2/c^2)} = l/c.$$

Classical relativity justifies the negative results obtained by Michelson and Morley in any referential and for any considered velocity of light, irrespective of the origin and the history of the employed light, therefore irrespective of its real velocity, either c or a different value resulting by its vector summing with the velocity of its source.

And this classical summing $\bar{c} \pm \bar{v}$ of the velocity of light $\bar{c}$ with that of its source $\bar{v}$ solves as simple several other similar experiments, also for a long time claimed as evidence for the relativistic invariance of the velocity of light:

"It may be noticed that the list does not include the most famous of all relativistic phenomena – the Michelson-Morley experiment. The reason is that the Ritz theory was relativistic (in the Galilean sense) and thus automatically explained the negative results of this experiment, as well as the Kennedy-Thorndike and Trouton-Noble experiments."[34]

The results of experiments concerning the frequency shifts of the light after its reflection on moving mirrors were also illicitly assigned to special relativity, because all these experiments can be very well solved classically, since in compliance with classical physics the initial velocity c of the light becomes $c(1 \pm 2v/c)$ after its normal reflection on a moving plane mirror whose uniform velocity $\bar{v}$ is perpendicular to its surface, and this change in velocity means a corresponding change in frequency of the reflected light, $\nu = \nu_0(1 \pm 2v/c)$, where ν_0 is the frequency of light before its reflection.

"In laboratory experiments with moving mirrors the frequency of the light reflected from an approaching mirror is $\nu(1 + 2\beta)$, where $v \equiv \beta c$ is the velocity of the mirror and ν the frequency of the light source. This is the result of observation and of special relativity. It is also the result of the theory under discussion."[34]

In these experiments relativity and classical physics are equally efficient, as long as both relativistic $\nu = \nu_0(1 + \cos\alpha \cdot v/c)/\sqrt{1 - v^2/c^2}$ and classical $\nu = \nu_0(1 + \cos\alpha \cdot v/c)$ equations of the Doppler effect become identical when $v \ll c$ and $v^2/c^2 \approx 0$. In these formulas ν is the frequency of a photon emitted under an angle α by an emitter with linear velocity v, while ν_0 is that of the same photon emitted by an emitter at rest.

But if the emitters have higher velocities and the factor $\sqrt{1 - v^2/c^2}$ cannot be neglected any longer, it appears a particular difference when the moving emitters radiate in a direction perpendicular to their direction of motion, $\alpha = \pi/2$: indeed, while in classical physics the Doppler effect is null in this case, $\nu = \nu_0$ when $\cos\alpha = 0$, according to the above relativistic formula there is a *transverse* Doppler effect even in these conditions,

$$\nu = \nu_0/\sqrt{1 - v^2/c^2},$$

and this difference can evidently decide the correct equation of the Doppler effect by detecting, or not, a frequency shift for the photons radiated under angles $\alpha = \pi/2$ by emitters with known linear velocity $\bar{v}$.

However, first experimental data invoked many decades as being a proof for the transverse Doppler effect were those found by Ives and Stil-

---

[34] J. G. Fox, *Evidence against Emission Theories*, Amer. J. Phys. **33**, (1965) 1.

well[35], who determined the frequency gap $\Delta v = v_{max} - v_{min}$ between the two maximum and minimum frequencies of the monochromatic photons emitted forward and, respectively, back by hydrogen atoms previously accelerated in electromagnetic field to a velocity $v \approx 10^6$ m/s, and noticed that the middle of this interval Δv does not coincide with the frequency v_0 recorded when the atoms radiate at relative rest, but with a little higher frequency $v = v_0(1 + 2 v^2/c^2)$, which corresponds to the relativistic transverse shift $v = v_0/\sqrt{1 - v^2/c^2}$ for small velocities $v \ll c$ of the radiating particles. And although only after a while these experimental data were claimed as experimental confirmation of the transverse Doppler effect, in the next decades their relativistic interpretation was reinforced by other researchers in different experimental versions.

Still such transverse frequency shifts of spectral lines emitted by atoms in motion has in fact a much simpler non-relativistic explanation, since conforming even to special relativity in hydrogen atoms electromagnetically accelerated to a certain velocity v the radiating electrons have an increased mass $m = m_0/\sqrt{1 - v^2/c^2} > m_0$, and consequently these moving atoms have a proportionally higher Rydberg constant $R = me^4/8\varepsilon_0^2 h^2 = R_0/\sqrt{1 - v^2/c^2}$, where ε_0 is permittivity of vacuum and $R_0 = m_0 e^4/8\varepsilon_0^2 h^2$ is the Rydberg constant of the atoms at rest. But an increased Rydberg constant means spectral frequencies increased exactly with the same factor $1/\sqrt{1 - v^2/c^2}$, this is an elementary consequence.

Therefore, in their experiment Yves and Stilwell recorded the frequency $v = v_0/\sqrt{1 - v^2/c^2} \approx v_0(1 + 2 v^2/c^2)$ at the middle of the frequency gap $\Delta v = v_{max} - v_{min}$ simply because all frequencies emitted by the moving H atoms increase proportionally to their increased Rydberg constant, in turn depending on the relatively small velocity v reached after their acceleration in electromagnetic field, and not owing to Einsteinian relativity of time.

In consequence, when the dependence on velocity $R = R_0/\sqrt{1 - v^2/c^2}$ of the Rydberg constant proper to the radiating H atoms previously accelerated in electromagnetic field is not forgotten any longer, the classical Doppler equation for the monochromatic photons radiated by these H atoms becomes $v = v_0(1 + \cos \alpha \cdot v/c)/\sqrt{1 - v^2/c^2}$, in full compliance with

---

[35] H. E. Ives and G. R. Stilwell, *An experimental study of the rate of a moving atomic clock. I*, J. Opt. Soc. Am. **28**, (1938) 215 and *II*, J. Opt. Soc. Am. **31**, (1941) 369.

experimental data obtained by Ives and Stilwell, while the relativistic equation of the Doppler effect becomes $v = v_0(1 + \cos\alpha \cdot v/c)/(1 - v^2/c^2)$, clearly invalidated in this experiment falsely claimed to be a proof of the transverse Doppler effect. Of course, this surprising turnaround acknowledges directly the classical dependence of the velocity of light on the velocity of the radiating hydrogen atoms.

(2) Other experiments claimed the same velocity of the light emitted in opposite directions by a macroscopic source. For example, De Sitter's measurements on binary stars were long time popularized in this regard, but finally their mentioning was abandoned owing to the justifiably raised objections. More recently, another experiment[36] was also largely invoked, since it would have confirmed the same velocity for the photons emitted from the two ends of the solar equator, where the emitters have linear velocities of about $4 \cdot 10^5$ m/s and contrary directions of motion because of the Sun's own rotation. Nevertheless, eventually its exhaustive critical analysis[37] restored the truth:

"We do not consider this work entirely convincing. On the one hand, the author begins with some a priori assumptions concerning the speed of light after reflection from a mirror; on the other hand, owing to the low accuracy inherent in the method, the observed results have a large scatter, several times larger than the effect expected from classical theory. To obtain the final result the author had to treat statistically a very large number (1727) of observations, in which very large deviations were excluded outright. In view of what has been said this work should not be regarded as a direct experimental verification of the independence of the speed of light on the source velocity."

(3) Two of the *classical* experimental tests of general relativity refer to the frequency shift of the light in gravitational fields with no change of its invariant velocity c. Still these two relativistic experimental tests have very simple classical solutions.

Thus, when monochromatic photons with frequency v are emitted on radial directions by a massive star with mass M and radius R, conforming

---

[36] M. A. Bonch-Bruevich and V. A. Molchanov, *A new optical relativistic experiment*, Opt. Spektrosk. **1**, (1956) 113. (in Russian)

[37] A. G. Baranov, *A Method to Verify Experimentally that the Speed of Light is Independent of the Velocity of the Source*, ZhETF **40**, (1961) 860. (in Russian)

to classical physics their initial speed c has a small diminution $\Delta v \ll c$ because of gravitational attraction exerted on them by the emitting star. Or, just because this diminution is negligible as compared to initial velocity c, in first approximation we can consider that the emitted photons go through the stellar gravitational field with a practically constant speed c, and consequently we can calculate firstly their average gravitational deceleration on the whole distance H from the star to the Earth by using the trivial theorem of mean value, $a_m = \frac{GM}{H} \int_R^{R+H} dr/r^2 = -GM/R(R+H)$, and then their total decreasing in speed $\Delta v = a_m t = -GM/Rc(1+R/H)$, where $t \approx H/c$ is the total travel time. Since at a big distance star-Earth $R/H \approx 0$ and $\Delta v = -GM/Rc$, in accordance with classical formula of Doppler effect $\Delta \nu/\nu = \Delta v/c$, this gravitational deceleration of the photons leads to a relative redshift equal to the relativistic redshift experimentally found as far back as in the 1920's for the monochromatic light emitted by Sun or Sirius,

$$\Delta \nu/\nu = \Delta v/c = -GM/Rc^2.$$

Likewise, when monochromatic photons with frequency ν and velocity c are emitted perpendicularly to the Earth's surface from a height H, conforming to Newtonian physics their velocity increases to $(c + gt)$, where g is the relatively constant gravitational potential on the Earth's surface and $t \approx H/c$ is the travel time (because $\Delta v = gt \ll c$), so that an observer at rest on the Earth's surface notices a relative blueshift of initial frequency ν,

$$\Delta \nu/\nu = \Delta v/c = gH/c^2.$$

This formula common for classical and special relativity was experimentally confirmed by Pound and Rebka[38] by using monochromatic γ-rays sent to the Earth from a height $H = 30$ m.

General relativity reaches the same formulas $\Delta \nu/\nu = GM/Rc^2$ and $\Delta \nu/\nu = gH/c^2$ based on the dependence of the time on the local gravitational potential, which excludes the change of the velocity of light in gravitational field. Very well, but why are these two experimental tests of general relativity presented always in books or treatises without mentioning their very simple classical solutions?

(4) The last experimental test of general relativity included in this critical review concerns the tangential passing of electromagnetic radiation to

---

[38] R. V. Pound and G. A. Rebka, *Gravitational redshift in nuclear resonance*, Phys. Rev. Lett. **3**, (1959) 439.

the solar globe, which had a historical key role in the acceptance of the whole theory of relativity.

Indeed, in 1911 Abraham doubted the new special relativity with a very simple argument: if astronomical observations confirm a curved trajectory of the light rays emitted by stars when they pass near to the solar globe before reaching the Earth (an effect already presumed for many decades), then the second postulate of special relativity is experimentally invalidated. In reply, Einstein assumed such a bending effect as a main target for his new theory of gravitation already initiated for some years, and after some not too clear attempts in this respect[39], he reach an angle $\alpha = 1".7$ of gravitational deflection of the light passing tangentially to the solar globe[40], twice larger than the value $\alpha = GM/Rc^2 = 0".87$ calculated by Newtonian mechanics, where G is the constant of universal attraction, M is the solar mass and R is the Sun's radius. But the Einstein's interpretation of this doubled gravitational effect was really amazing:

"It may be added that, according to the theory, half of this deflection is produced by the Newtonian field of attraction, and the other half by the geometrical modification ("curvature") of the space caused by the Sun."[41]

Anyway, as nobody objected that any bending of light due to Newtonian gravitational force means implicitly a corresponding change in the speed of light, even if it is only 50 % responsible for the whole deviation angle!, or why this mixture between Newtonian theory of gravity and supplementary relativistic effects have not been considered in all the other experimental tests of general relativity, for example the frequency shifts in gravitational fields, finally what remained was just the new Einsteinian vision about the four-dimensional space-time curved by the mass. As a matter of fact, in those years some physicists even expected a new theory of gravitation able to solve at last the older controversy on the Mercury's perihelion precession, very hastily considered by them to be classically unsolvable.

Since in these circumstances this scientific struggle became so important, several measurements by a complicated photographic method were

---

[39] C. Lane Poor, *The Relativity Deflection of Light*, J. Roy. Astron. Soc. Canada **XXI**, (1927) 225.

[40] A. Einstein, *Die Grundlage der allgemeinen Relativitätstheorie*, Ann. Phys. **354**, (1916) 769.

[41] A. Einstein, *Relativity: The Special and General Theory*, H. Holt, New York, 1920.

carried out quickly enough in spite of all difficulties inherent at that time, but the more officious backers of the new Einsteinian vision popularized exclusively and in a spectacular manner the angle $\alpha = 1".65 \pm 0".45$ found by Eddington and Kottingham in 1919, by which they change Einstein into a Magus of science, although exactly in the same year Kromelen and Davidsson measured in better technical conditions $\alpha = 1".98 \pm 0".12$ (value increased later by other researchers to $2".05$, $2".07$ and $2".16$ only by remaking the calculi), and in the next decades many other researchers found deviation angles from $0".9$ to about $3"$ by the same method.

Moreover, all the reviewing works of the specialists in the field have revealed the big lack of precision of the used photographic method, due to numerous and hardly detectable technical factors which misstate on photographic plates the real positions of the stars in relation to the Sun, so that the same photos can lead to different angles of deviation if they are processed by different calculation methods, or by person with different convictions and even interests, for which reasons all the known measurements can only confirm a curved trajectory at passing in the Sun's closeness, but not at all a doubtless deviation angle able to decide between the two theories of gravitation. Here is a lucid conclusion[42] after a critical review of all these experiments using the photographic method:

"In 1924 Professor Esclangon, the director of the Paris Observatory, wrote: The observations neither confirm, nor contradict the Einstein law of deviation. They only indicate, if all presumptions about systematic errors are excluded, the deviation near the Sun, but without determining the law and without the exact value of the deviation at the edge of the Sun.

I am afraid that now, after thirty-five years and six successful observations of the Einstein effect, the same rather skeptical words can be repeated again."

More, although absolutely all of experimental values obtained by photographic method are placed much over the classical value $\alpha = 0".87$, and thus they seem to support the relativistic angle $\alpha = 1".74$, in the second half of the past century many researchers have brought forward a series of factors capable of amplifying the deviation angle in solar atmosphere. Two of them, the refractive effects within the solar atmosphere and the influence of

---

[42] A. A. Mikhailov, *The Deflection of Light by the Gravitational Field of the Sun*, MNRAS **119**, (1959) 593.

the solar magnetic field, can certainly change in a more or less measure the trajectory of all kinds of electromagnetic radiation, either luminous rays, radio waves or microwaves. And among these two phenomena potentially able to influence the deflection angle of electromagnetic radiation in the solar atmosphere, refraction is undoubtedly the most important. Thus, from observations about the scattering of the radio waves at their passing through the solar trap one deduces that the refraction effects can be significant up to a distance of ten solar radii[43], even if at these distances the solar atmosphere is much rarefied.

For all that, although refraction effects are taken into account by special relativity even in intergalactic spaces, by far the nearest to a perfect vacuum, they have always been ignored as regards their influence on the velocity of the light propagating in the much denser solar atmosphere, even the specialists in the field[44] could not find a place in physics journals.

Also, although the action of the magnetic field upon the photons has not been investigated up to the level of the very fine effects, there are yet some indicia in this regard, as the known rotation of polarization plane in strong magnetic fields, noticed by Hertz since the 19th century, or different modifications of spectral lines in very intense optic fields[45].

More, both the density and magnetic induction of the solar atmosphere vary locally between very large limits (up to thousands times) due to the huge radial convection fluxes in photosphere, and these continuous variations could justify the wide dispersion of experimental results, in addition to the acknowledged imprecision of the photographic method (some statistical surveys even seem to exhibit a certain dependence of the deviation angle on the cycles of solar activity).

Another possible factor liable for this large dispersion of experimental results could be the Sun's non-uniform density, higher in its equatorial zone as against the density at the poles, as the faster rotation of this zone proves.

---

[43] R. H. Dicke, *The enigmatic periodicity of the solar oblateness*, Solar Physics **37**, (1974) 271.

[44] L. Wahlin, *The bending of light by gravity*, Colutron Research Corporation (http://www.colutron.com/download_files/Bend.pdf).

[45] J.P. Barrat and C. Cohen-Tannoudji, *Enlargement and displacement of magnetic resonance radiation caused by an optical excitation*, J. Phys. Radium **22**, (1961) 443.

A. Kastler, *Displacement of energy levels of atoms by light,* J. Opt. Soc. Am. **53**, (1963) 902.

Of course, a non-uniform density of the solar mass means different gravitational potentials on the Sun's surface, therefore different angles of deflection depending on the distance between the passing points of the luminous rays and the Sun's equator.

A century ago first experimental data concerning the deflection angle of the luminous rays tangent to the solar globe aroused much enthusiasm, but then gradually diminished by some critical reviews of the less hotheaded specialists in the field. However, in the last decades, a new method using radio waves[46] seems to tip definitively the balance in favor of relativistic angle 1".74, by experimental results that confirm it with an impressive accuracy. But an exhaustive critical analysis of data supplied by this method is still expected from those able to quantify at the present level of knowledge the refraction effects of radio waves within the solar atmosphere, all the more so as in this regard we have already the negative experience with the older photographic method.

Or, as long as all possible mechanisms able to increase the deflection angle in the solar atmosphere are not exhaustively identified and quantified as accurate as possible, the real size of gravitational deflection angle in vacuum remains a still open problem.

Experimental test in reality experimental denial

The last relativistic test referring to electromagnetic radiation passing tangentially to the solar globe was proposed in 1964 [47], and consists in the longer time travel of radio waves on the route Earth—Mercury (or Venus) — nus) — Earth in case they pass very near to the Sun, in comparison with the normal duration of this travel. Later similar retardations were also noticed to radio waves emitted by different cosmic sources.

Conforming to general relativity these delays are determined by the time dilation in a stronger gravitational field, together the longer travel of the waves on the geodesic lines of the four-dimensional space curved near the Sun, but again we can easily explain this effect of travel delays in classical context.

---

[46] K. J. Treschman, *Recent astronomical tests of general relativity*, Int. J. Phys.Sci. **10**, (2015) 90.

[47] I. I. Shapiro, *Fourth test of general relativity*, Phys. Rev. Lett. **13** (1964), 789.

Indeed, according to classical physics such delays can appear for at least two reasons. For example, a temporary decrease in velocity certainly appears inside the dense solar atmosphere with refractive index n, wherein the speed of electromagnetic radiation $v_\gamma = c/n$ can be significantly smaller than c, and this smaller velocity in the nearness of the solar globe means evidently a delay of their travel in comparison with the normal route. But beside this temporary decrease in velocity by refraction inside the solar atmosphere, the speed of the photons passing very close to the Sun must also decrease definitively, as a consequence of their larger average distance to the solar globe when they approach the Sun, as against the average distance during their moving away from the Sun. In other words, the total acceleration of the photons while approaching the Sun is a little smaller than their subsequent deceleration, what means evidently a smaller final velocity of the waves as against their initial velocity c.

Unfortunately, until now no one was interested in an exhaustive study about these influences in classical context, unlike the massive attention given to Shapiro's relativistic calculations. However, an experiment carried out soon after Shapiro's work can make the difference.

Very curiously, long time no one of those involved in researching the passage of electromagnetic radiation through the solar gravitational field did not seek to check for a possible definitive change of its frequency after this tangential passage, which in principle could still be easily found no matter how small it is thanks to the high precision of spectroscopic measurements. This lack of interest was at least strange, because such a trivial check could preclude any doubt on the invariance of the speed of light even in the solar gravitational field, as long as, differently from the relativistic predictions in these questions, in classical physics the final velocity change Δv inevitably experienced by the radio waves inside the Sun's gravitational field remains permanent even after they leave the strong gravitational field of the Sun, and consequently, besides a little longer total travel time of the radio waves when their trajectory is more or less tangent to the solar globe, their frequency has also to have a definitive decrease $\Delta \nu$ conforming to the known formula $\Delta \nu / \nu = \Delta v / c$.

However, such a measurement was done yet for the microwaves $\lambda = 0.212$ m ($\nu = 1.420406 \cdot 10^8$ Hz) coming from Taurus constellation after their passing through the gravitational field of the solar globe, when a very small decrease $\Delta\nu \approx 150$ Hz in their frequency was found[48], which proves a similar effect also in the case of the radio waves passed through the solar gravitational field.

Or, conforming to general relativity only an observer placed on the solar globe could observe a frequency change, namely a blueshift caused by a slower elapsing of the time on the solar surface, but an observer placed on terrestrial surface has to notice exactly the initial frequency, no matter if previously these microwaves passed or not through one or several gravitational fields with higher potential. Therefore, such a decreasing in frequency of an electromagnetic radiation recorded on the Earth after passing through the Sun's gravitational field, simultaneously with an increase of its travel time, can appear only if the speed itself of this radiation decreases after its passing through the solar gravitational field, another reliable explanation simply cannot exist.

At its time, this experimentally noticed decreasing in frequency was euphemistically declared "inexplicable for general relativity" and then gradually forgotten along with some few *ad-hoc* revelations hastily fabricated to exculpate general relativity. In reality, if an electromagnetic radiation exhibits a definitive frequency shift after its very short passing through a stronger gravitational field, general relativity is definitively denied along with the whole theory of relativity, such an "inexplicable" reality is a conclusive denial enough for such a generalized conclusion.

Postulate for an imaginary space

Certainly the main reason of the stubborn adversity roused from the very outset by special relativity was its second postulate on the invariant velocity of light, the starting point of the whole theory. According to this postulate a ray of light has always the same speed against an observer, no matter if the latter moves towards it or in the same direction. Even if the observer's velocity would be very near to the velocity $c = 3 \cdot 10^8$ m/s of the light, say $v = 0.999\,c$, the speed of light measured by him remains $c = 3 \cdot 10^8$ m/s, no

---

[48] D. Sadeh, S. H. Knowles and B. S. Yaplee, *Search for a Frequency Shift of the 21-Centimeter Line from Taurus A near Occultation by Sun*, Science **159**, (1968) 307.

D. Sadeh, S. H. Knowles and B. Au, *The Effect of Mass on Frequency*, Science **161**, (1968) 567.

matter what directions of motion the observer and the light have, from parallel to opposite. Obviously, this postulate contravenes flagrantly the human knowledge founded on Galilean relativity.

In the last century this hypothesis concerning the unique role of the light in Universe has been very much discussed in all respects, including cases when the second relativistic postulate was directly invalidated by the measured velocities of the different kinds of electromagnetic radiation, but never the logical consequences of its validity limited to the emptiness. In 1905 this limitation was obligatory, whereas half a century before Fizeau measured in the water a velocity of light equal to $0.75\,c$, but because after some few decades it was ascertained that in water, for instance, the protons or the electrons can have velocities even over $0.98\,c$, therefore much higher than the speed of light, a question ought to have been asked: in water, and generally in any not void space where the light can have a velocity smaller than other elementary particles with rest mass, is special relativity valid any longer?[49]

This question is more than justified, since the essence of special relativity is the so-called Einsteinian simultaneity of events, whence it results the Einsteinian relativity of time and space on the basis of which all phenomena claimed to be relativistic effects are solved, from Michelson-Morley experiment to different kinds of gravitational Doppler effects. Or, this Einsteinian simultaneity of events can be demonstrated only from the hypothesis that the velocity of light is both invariant and the highest in Universe, intangible for all the other physical entities. Moreover, the invariance of the speed of light and its presumed quality to be the fastest signal in Universe are inseparable attributes in special relativity, each of them can be derived from the other by a logical reasoning.

In consequence, in a physical space where other physical signals are faster than the light, and thus they could induce causal interactions even in situations where the slower optical signals cannot do it, the Einsteinian simultaneity of events cannot be demonstrated any longer, and so special relativity remains without its motivational basis. In a physical space where the electrons with rest mass reach velocities higher than the photons without rest mass, for instance in water, the velocity of light loses its magic status of

---

[49] Really amazing, many decades the Fizeau's experiment was claimed as an experimental proof of special relativity!

absolute limit in the whole Universe, and special relativity loses its indispensable prerequisite.

But not only the translucent spaces on the Earth, from water to terrestrial atmosphere, are physical spaces where special relativity becomes a superfluous theory, such a situation had to be admitted even for intergalactic spaces, considered to be in nature the nearest to a perfect vacuum. Indeed, any radio impulse of a pulsar is first recorded by a radio receiver tuned on high frequencies, and only after that the lower frequencies of the same signal are gradually received. Or, this lateness proves clearly different velocities of electromagnetic radiation depending on its frequency: close to c for the high-frequency radio waves, but more and more inferior to c for those having lower and lower frequencies. And these different velocities of radio waves with different frequencies are explained by the dependence of refraction index of electromagnetic radiation on its frequency.

Therefore, special relativity has in view refraction effects for electromagnetic radiation that travels through intergalactic spaces, by far the closest to an absolute vacuum, but ignores them in the case of the light crossing the much denser solar atmosphere. So what?

But not this relativistic duplicity is important here, what matters primarily is this implicit acknowledgement of the fact that in the real world the speed of light is nowhere invariant. In consequence, if anywhere in Universe the velocity of light varies depending on a certain physical parameter, for example its frequency, how can be accepted a theory whose validity is strictly conditioned by an absolutely invariant velocity of light, as long as this absolute invariance can be accepted nowhere in Universe?

The dark matter more recently discovered in the whole Universe raises similar problems, since even the last experiments for finding its spatial distribution prove a weak interaction of it with electromagnetic interaction, one way, or other. And even only the certitude of its existence entirely change the meaning of what was called "vacuum" until now, a syntagm essential for special relativity. Or, if no physical space in Universe is absolutely passive for electromagnetic radiation, special relativity can be at most a beautiful theory for an imaginary space perfectly void of substance, but not at all a theory worthy to be considered in the real world.

SPINNING RING ELECTRON

Brief history

Soon after J. J. Thomson proved experimentally in 1897 that the electron is not only an undefined charge of electricity, but also a particle with determinate mass, more and more experimental data suggested that the electron is the elementary source of a dipole magnetic field with axial symmetry, similar to those generated at macroscopic level by circular electric currents. First reference to a rotating electron seems to have been done by Schwarzenschild[50] in 1903, and in 1907 Stark[51] wrote about the axial symmetry of the inner structure of the electron:

"It is therefore justified to assign to the electron, as the elementary carrier of energy and electric charge, the universal constant h_0 and also an elementary time T_0.

The fact that we assign to the electron a proper period T_0, and thus an elementary angular velocity $w_0 = 2\pi/T_0$, leads us to believe that a cyclic movement is centered within the electron. If this is correct, the electron must have an anisotropic structure. We want to present the hypothesis that this cyclical movement has a circular form, with a linear axis perpendicular on the motion plane at its central point."

But the first well defined model of spinning ring electron was outlined by Parson[24]:

"The essential assumption of this theory is that the electron is itself magnetic, having in addition to its negative charge the properties of a circuit current whose radius (finally estimated to be 1.5×10^{-9} cm) is less than that of the atom but of the same order of magnitude. Hence it will usually be spoken of as *the magneton*. It may be pictured by supposing that the unit negative charge is distributed continuously around a ring which rotates on its axis

---

[50] K. Schwarzenschild, *Über die Bewegung des Electrons*, Nachr. Ges. Wiss. Göttingen, Math. Phys. **5**, (1903) 245.

[51] J. Stark, *Elementarquantum der Energie, Modell der negativen und positiven Elektrizität*, Phys. Zeitschr. **8**, (1907) 881.

(with a peripheral velocity of the order of that of light); and presumably the ring is exceedingly thin."

Three years later Compton[52] confirmed this picture by his experimental data on electron scattering, and in addition brought in the concept of the dimensional "flexibility" of the electron, because the size of this simultaneously elementary particle and quantum of electricity appeared to vary under different experimental conditions. Also, in a meeting of Physical Society of London in 1919 other physicists noted that a series of other experimental data support, or even demand, the structural model of spinning ring electron.

But yet, despite these favorable developments, in the next decade the model of ring electron was abandoned quickly enough, and a first reproach has been just the spatial extension of its elementary electric charge, physically impossible because of the repulsive electromagnetic forces acting between all its constituent parts, which would not allow their aggregation as a stable spatial structure (although neither the pointlike quantum of electricity has been a concept exceedingly clear!). Still the main reason of this very early abandonment of the ring electron model was its incompatibility with other recent branches of physics, as the Bohr's atomic theory and the Einstein's special relativity.

For example, a decisive argument against the spinning ring electron was the angular momentum $L_e = h/4\pi = 0.527 \cdot 10^{-37}$ J·s postulated for electron by Goudsmit and Uhlenbeck[53] in their already mentioned theory on the fine splitting of hydrogen spectral terms, a theory suggested by Pauli and quickly accepted by Bohr in order to get rid at last of this spectral splitting still unsolved and consequently very disturbing for his atomic model. Or, such a spin $L_e = h/4\pi$ of the electron means implicitly a gyromagnetic ratio $g_e = M_e/L_e = e/m_0$ of the particle, twice larger than that one of a classical ring electron assimilated to a very small circular electric current, $g_e = e/2m_0$. Therefore, as Goudsmit and Uhlenbeck asserted explicitly that the electron could not be a classical spinning object, by accepting their theory Bohr accepted explicitly to shift the electron in a world ruled by laws entire-

---

[52] A. H. Compton, *The Size and Shape of the Electron*, Phys. Rev. II, **14**, (1919) 247.

[53] G. E. Uhlenbeck and S. Goudsmith, *Elementarquantum der Energie, Modell der negativen und der positiven Elektrizität*, Naturwiss. **13**, (1925) 953 and *Spinning Electrons and the Structure of Spectra*, Nature **117**, (1926) 264.

ly different from those acting at macroscopic level, and this profound scission became quickly the creed of the new quantum physics.

Besides the new atomists, another very active and influential group was even more interested in ruling out the concept of classical rotating electron: Einstein and his adherents. Indeed, special relativity is a theory of the pointlike electron exclusively, a theory definitely inadaptable to a spatial electron with a certain inner structure.

In the subsequent years this new concept of electron with no internal structure, but with kinetic and magnetic moments, was gradually consolidated on two planes.

On the one hand, all experimental data known until then to be consonant with a classical gyromagnetic ratio $g_e = e/2m_0$ of the electron were contested some way or other, including those obtained by Einstein and de Haas in their famous experiments begun in 1915, whose results were gently explained and exculpated by a too high wish of the two physicists to demonstrate the real existence in substance of the "molecular" circular currents presumed by Ampère long before.

On the other hand, the discovery of the positron in 1932 was used for a positive reappraisal of the relativistic equation of the electron found out by Dirac in 1928, but ignored for a while owing to its oddities. Actually the Dirac's purpose was to write the Lorentz invariant equations for the movement of the free electrons and protons, according to which the internal properties of these particles had to remain the same for observers both in motion or at rest. But the most eccentric "peculiarity" of Dirac's relativistic equation has been the foreknowledge of a domain of negative energies $E < -m_0c^2$ of the electrons, separated by an interval $\Delta E = 2m_0c^2$ to that one of positive energies $E > +m_0c^2$ considered until then. While the latter belongs to *usual* electrons, with positive mass and energy, the new domain would belong to some *unusual* electrons, with negative mass and energy. These *unusual* particles should have had properties contrary to our experience, e.g., to move in an opposite direction to the forces applied to them! More, Dirac refused vehemently to rule out the negative domain $E < 0$ as an unphysical solution of his equation, and even alleged obstinately that the proton is in fact an *unusual* electron. However, after discovering the positron, which was claimed as a brilliant acknowledgment of Dirac's predictions, the interest for the relativistic equation of the electron was revived (the proton was gradually forgotten, because all predictions regarding it failed lamentably), and consequently important efforts were engaged for the sake of providing a *natural* interpretation for Dirac's mathematical results. And indeed, after "long-lasting struggles" a theory just as strange was devised, according to which an uniform stock of *unusual* electrons exists somewhere in an *unnoticeable* parallel world, but they can be converted to normal, noticeable particles if an energy greater than $2m_0c^2$ is given to them by undefined means. Still more, for legitimizing these transformations even an *unusual* algebra was specially devised. And so the two moments of the electron became *relativistic quantum effects* only apparently similar to those classically defined, but in fact with no connection with a rotation motion of a composite electron. Here is Purcell[54] explaining much later this self-deceit:

"There is no point in trying to devise a classical model of this object; its properties are essentially quantum mechanical. We need not even go so far as to say it *is* a current loop. What matters is only that it behaves like one in the following aspects: (i) it produces a magnetic field which, at a distance, is that of a magnetic dipole; (ii) in an external field B it experiences a torque equal to that which would act on a current loop of equivalence dipole moment; (iii) within the source, $div\ \boldsymbol{B} = 0$ everywhere, as in the ordinary sources of magnetic field with which we are already familiar."

Of course, it is quite useless to ask what "within" can a point have, or how can a pointlike object experience a torque involving two distinct forces acting simultaneously at two different points of a body, such childish questions cannot concern a Nobel prize laureate.

Still Dirac's theory did not resist much time, because in 1947 Lamb discovered the frequency gap $\Delta v = 1.058 \cdot 10^9$ Hz between the hydrogen spectral terms $2S$ and $2P_{1/2}$ predicted to be identical by Dirac, and Kusch measured by magnetic resonance a magnetic moment of the free electron a little larger than the Bohr magneton value resulted from Dirac's relativistic equation of the electron. Very strangely, both Lamb and Kusch received a Nobel prize in 1955 for discoveries that have invalidated the Dirac equation rewarded with the Nobel prize in 1933!

Although Dirac's theory is presented even today in treatises as a fundamental work of quantum relativistic physics, its experimental invalidation

---

[54] E. M. Purcell, *Electricity and Magnetism*, Berkeley Physics Course Vol. II, McGraw-Hill, New York, 1965.

brought forth a state of discomfort, later amplified by other troubles appeared meanwhile. For instance, conforming to a basic principle of the new quantum electrodynamics, the mass and energy of an electrically charged particle depend inversely proportional on its size, and consequently the pointlike electron ought to have an infinite mass and energy. That is why, in the next decades it was gradually developed a so-called theory of renormalization, initially a pure mathematical procedure doubted by other mathematicians for some not too orthodox operations, and then even by its formulators, yet ultimately claimed as a way able to reconcile the concept of the pointlike electron with its known mass, charge and moments, whose most recent values have been exactly calculated after an infinite series of progressively small corrections. And the other pointlike leptons have been entirely forgotten!

In their already mentioned work[32], Ekström and Wineland pass briefly in review the three main stages in understanding the electron:

"Viewing the electron as a rigid, rotating body is somewhat naive; after all, the motion of the particle must be described by the laws of quantum mechanics, where the notions of size and velocity cannot even be defined beyond a certain level of precision. Indeed, the model has grave flaws, some of which recognized only days after it was proposed. For example, it turns out that the rotational velocity at the surface of the electron is greater than the speed of light. Another source of difficulty is the size attributed to the electron. The mass or energy of an electrically charged particle depends inversely on its size. One reason this is so can be understood by noting that energy is required in order to pack the repulsive negative charge of the electron into a finite volume. The smaller the volume, the larger the energy needed. According to this scheme, the quite small mass or energy of the electron implies that it should have a rather large size. Experiments in which electrons are scattered by other particles, however, effectively measure the size of the electron, and they indicate that the radius must be exceedingly small. Indeed, all experimental data gathered so far are consistent with the idea that the electron is a point particle, entirely without extension. The arguments presented here then predict that the electron mass is infinite, a manifest absurdity.

Still another reason for doubting the accuracy of the mechanical model, and that of the Dirac theory as well, derives from refined measurements of the g factor of the electron. Experimental evidence has shown that g is not exactly 2 but rather is greater than 2 by about .1 percent; in other words, its value is roughly 2.002. The Dirac theory could not accommodate such an adjustment.

In the 1940's these problems were resolved by abandoning the mechanical model of the electron and devising a new and more abstract theory, quantum electrodynamics. In quantum electrodynamics the electron is allowed to be a dimensionless point particle and its mass is allowed to be infinite, at least in principle. Surprisingly, there is a mathematical procedure (called renormalization) that cancels this infinity and recovers the observed, finite properties of the electron, including the g factor.

It is far from obvious how a particle with zero radius can have spin angular momentum or magnetic moment. If quantum electrodynamics offers no consistent mental picture, however, it does provide an explicit procedure for calculating the numerical values of the various properties of the electron."

Still this renormalization procedure has not cleared all problems raised by the pointlike electron. For example, if the infinite mass of the pointlike electron is a recognized absurdity removed by renormalization, is its infinite mass density more acceptable?

And concerning the dipole magnetic field generated by the electron in the surrounding space, the essential problem is not the small difference between its size measured for the free electron and that determined from the line spectrum of hydrogen on the basis of atomic theory, but above all its physical source, whose absence in the case of the quantum relativistic pointlike electron cannot be conjured away by that word "intrinsic" devoid of any physical sense.

Also, all the above flaws imputed to a generic "mechanical" model of rotating electron are entirely gratuitous in the concrete case of the spinning ring electron:

(1) The spinning ring electron has never peripheral velocities greater than the speed of light, its spin is immutable, while its size and magnetic moment depend permanently and within wide limits on the external magnetic induction in accordance with the basic laws of physics;

(2) As long as the spinning ring electron is an entity with a certain inner structure, therefore with determinate dimensions, any charge regarding

an absurd infinite mass of it is groundless. On the contrary, just the pointlike electron of quantum relativistic physics is in this position;

(3) As for the so-called *anomalous* magnetic moment of the electron and the very small dimensions found for the high-energy electrons scattered from slow nucleons or electrons, both questions are thoroughly explainable for the classical model of spinning electron and their common clue is the dimensional change experienced in very strong magnetic field by any macroscopic or infinitesimal circular electric current propagating in vacuum.

Thus, we have to take into account that the *anomalous* magnetic moment of the electron $M_e = 9.285 \cdot 10^{-24}$ A·m² is in fact that of the free electrons in a space with negligible magnetic induction, while the Bohr magneton $M_B = 9.273 \cdot 10^{-24}$ A·m² is calculated from the hydrogen line spectrum and therefore it is that of the atomic electron, which experiences in the fundamental atomic state an appreciable intra-atomic magnetic induction. Or, the spinning ring electrons cannot have an intrinsic, immutable magnetic moment, as the non-dimensional pointlike electrons, on the contrary, their magnetic moment varies depending on their radius, in turn dependent on the external magnetic induction, so that the very small difference between M_e and M_B is quite explainable for the spinning ring electron with changeable internal structure. And the very small dimensions found for electrons in scattering experiments have a similar cause, because these dimensions are those of the electrons at the moments of their collisions with the target magnetic particles, in the close proximity of which the incident fast electrons experience for a very short time a huge magnetic induction. Therefore, if the incident electrons are indeed infinitesimal electric currents propagating in vacuum, at the very moments when their motion directions really change by collisions, their radii have to be entirely different from their radius had in spaces with small, negligible magnetic induction.

But more about this dependence of the size and magnetic moment on the local magnetic induction after detailing the inner structure of the spinning ring electron.

Ring electrons as infinitesimal circular electric currents

If the spinning ring electron at rest has its mass $m_0 = 9.109 \cdot 10^{-31}$ kg and electric charge $e = 1.602 \cdot 10^{-19}$ C uniformly distributed on a subquantum orbit with radius r_e, its very fast rotation with linear peripheral velocity $c = 2.998 \cdot 10^8$ m/s around its symmetry axis is defined by angular momentum $L_e = m_0 c r_e$ and generates a dipole magnetic field with axial symmetry defined by magnetic moment $M_e = e c r_e/2$. These two moments give a gyromagnetic ratio of the spinning ring electron $g_e = M_e/L_e = e/2m_0$, which is identical, as normal, with that one of any macroscopic circular current. And because the electron magnetic moment $M_e = 9.285 \cdot 10^{-24}$ A·m² is known very accurately from experiments of electronic magnetic resonance, it results an accurate radius $r_e = 2M_e/ec = 3.866 \cdot 10^{-13}$ m of the ring electron, and an accurate angular momentum (or spin) of this particle $L_e = m_0 c r_e = 1.055 \cdot 10^{-34}$ J·s $= h/\pi$, where $h = 3.313 \cdot 10^{-34}$ J·s is the correct value of the Planck's constant.

Since the ring electron cannot be seen as a homogeneous annular continuum, such a vision would not be less absurd than the non-dimensional pointlike electron, this composite electron is to have its own distinct place somewhere on dimensional scale intuited by Newton three centuries ago:

"Now the smallest of particles of matter may cohere by the strongest attractions, and compose bigger particles of weaker virtue; and many of these may cohere and compose higher particles of weaker virtue; and many of these may cohere and compose bigger particles whose virtue is still weaker and so on for diverse successions, until the progression ends in the biggest particles on which the operations of chemistry and the colors of natural bodies depend, and which by cohering compose bodies of a sensible magnitude."[55]

From all accounts, five structural levels have been identified until now:

(1) Molecules, the smallest entities preserving the properties of substances;

(2) Atoms with dimensions measured in 10^{-10} m units;

(3) Atomic nuclei with dimensions measured in 10^{-14} m units;

(4) Nucleons with dimensions measured in 10^{-15} m units;

(5) Quarks constituent of nucleons, described together the photons and leptons as pointlike particles with no inner structure or anyway with di-

---

[55] I. Newton, *Opticks: or, a Treatise of the Reflections, Refractions, Inflections, and Colours of Light*, Book II, W.Innys, London, 1730.

mensions smaller than 10^{-18} m ... 10^{-19} m (although "with no inner structure" and "smaller than ..." are not at all the same thing!).

Therefore, for sketching the internal structure of the spinning ring electron a sixth level should be identified. But what entities could exist in a microcosm more profound than the elementary particles world?

Fortunately such an apparently risky plunge into a microcosm of our microcosm has been facilitated just by the more and more extensive table of elementary particles. Indeed, after a time the number of elementary particles increased so much, that their unexpected diversity seemed more and more to be rather combinational than intrinsic, as initially was thought. Moreover, this hypothesis has been strongly reinforced by the endless chain of the mutual transformations of all elementary particles. For instance, the electrons emit and absorb photons and neutrinos, they can even change into photons by annihilation or can synthesize hadrons by collisions at very high energy, but they also result as final products when some unstable particles disintegrate or annihilate, or can be created from photons having the necessary energy, and so on, and so on. All these realities have a so clear significance that even Heisenberg, one of the most vehement opponents of the concept of elementary particle with internal structure, wrote in 1960 [56]:

"Actually the experiments have shown the complete mutability of matter. All the elementary particles can be transformed, at sufficiently high energies, into other particles, or they can simply be created from kinetic energy and can be annihilated into energy, for instance, into radiation. Therefore, we have here actually the final proof for the unity of matter. All the elementary particles are made of the same substance, which we may call energy or universal matter; they are just different forms in which matter can appear."

And many other physicists have shared the same conclusion:

"From the general interdependence of particles it follows that each elementary particle consists to a certain degree of all the others, that is, all of them consist in their essence of something unique, a kind of a general basic material. Possibly in a not too far future physics will be able to define this basic material and construct out of it all the known particles, with all their properties."[31]

---

[56] W. Heisenberg, *Physics & Philosophy*, Allen and Unwin, London, 1959.

Certainly the simplest variant in this regard is to consider all elementary particles detectable by our experimental means as made of identical subparticles, undetectable for us just owing their exceedingly small size and mass. Conventionally we can call *preons* these universal subparticles existing in a far microcosm, a term already used in physics[57], but also in other sciences, for example in biology, in order to name primordial pre-entities able to bind to each other, and to form thus more or less complex structures.

In these circumstances, the spinning ring electron itself has to be a first preonic structure consisting in a variable number of preons equidistantly spread on its subquantum circumference, where all of them move constantly with the same linear velocity, equal to c when the electron is at relative rest. And in the absence of a distinct subquantum entity able to generate a central force responsible for such an uniformly circular motion of the preons in the subquantum space of the ring electron, we can only consider this circular subquantum motion as being a cyclotron motion, by seeing the rotating ring electron as an infinitesimal replica of an accumulating ring of electrons in uniform magnetic field, made at macroscopic level of monoenergetic electrons evenly spread on a circular trajectory where they experience any minute a constant magnetic induction perpendicular on their linear velocity. Therefore, in order to bring this similitude through, we are obliged to deem the preons as being in turn magnetic dipoles, most probably spinning ring particles too, as the simplest form of a stable structure able to generate a dipole magnetic field by its rotation around a proper axis.

If so, these magnetic preons should have not only determined mass m_p, but also determined angular momentum $\bar{L}_p$ and magnetic moment $\bar{M}_p$, whose orientation in the subquantum space of the ring electron can be either parallel or antiparallel to the moments $\bar{L}_e$ and $\bar{M}_e$ proper to the rotation motion of the whole ring electron, in accordance with the general rules of electromagnetism[58]. Or, these coplanar ring preons, with magnetic moments perpendicular to the orbital plane and parallel between them, and therefore

---

[57] J. C. Pati, A. Salam and J. Strathdee, *Are quarks composite?*, Phys. Lett. B **59**, (1975) 265.

[58] Logically the two possible orientations $\bar{L}_p \uparrow\uparrow \bar{L}_e$ and $\bar{L}_p \uparrow\downarrow \bar{L}_e$ should become a valid criterion for differentiating specifically the ring electrons from the ring positrons. And indeed, this difference will prove to be essential when the ring electron-positron pairs annihilate and create, or when the hadrons are created from high-energy ring electrons and positrons.

permanently equally distanced each to other on the subquantum orbit because of their repulsive magnetic interactions, can themselves provide that constant magnetic induction $\bar{B}_e$ necessary for each magnetic preon for its cyclotron motion on the subquantum orbit, induction evidently perpendicular to the orbital plane and the same, as absolute value and orientation, for all the preons.

An important question is certainly the motion equation of the magnetic preons on their subquantum orbit.

If we see an accumulating ring made of a big number n of slow electrons equally distanced on their cyclotron orbit with radius r as a circular electric current propagating in vacuum, the angular momentum of this stationary current is $L = nm_0v$, where v is the linear velocity of the electrons, and the dipole magnetic field generated by this current has a magnetic moment is $M = nevr/2$, wherefrom it results the gyromagnetic ratio of this macroscopic circular current, $g = M/L = e/2m_0$.

Also, if we see the spinning ring electron as an infinitesimal circular current with radius r_e, whose magnetic moment $M_e = ecr_e/2$ and spin $L_e = m_0cr_e$ give an identical gyromagnetic ratio $g_e = e/2m_0$, this equality $g = g_e$ between two gyromagnetic ratios of two stationary circular currents existing at macroscopic level and, respectively, in the subquantum space of the ring electron entitles us to consider this formula $g = g_e = e/2m_0$ as a constant physical quantity, valid at any dimensional level for the stationary circular currents. If so, we are entitled to extend this general validity to the known equation of cyclotron motion $r = m_0v/eB$, whose form in the subquantum space of the electron becomes $r_e = m_0c/eB_e$, where B_e is the constant magnetic induction felt by all magnetic preons on their subquantum cyclotron orbit with radius r_e. From the last formula it results $B_e = m_0c/er_e = 4.409 \cdot 10^9$ T.

More, when the ring electron is at rest in a uniform magnetic field $\bar{B}$, wherein its magnetic moment $\bar{M}_e$ can be only parallel or antiparallel to $\bar{B}$, and consequently the subquantum magnetic induction $\bar{B}_e$ can also be only parallel or antiparallel to $\bar{B}$, its radius $r_{e(B)}$ in this field decreases or increases conforming to the formula $r_{e(B)} = m_0c/e(B_e \pm B) = r_e/(1 \pm B/B_e)$, where the signs (+) and (−) correspond to the two opposite orientations possible in the external magnetic field, $\bar{B}_e \uparrow\uparrow \bar{B}$ and, respectively, $\bar{B}_e \downarrow\uparrow \bar{B}$. And because the inner peripheral velocity c of the spinning ring electron is not altered by the external magnetic induction B, exactly as in the case of macroscopic accumulating rings, the new radius $r_{e(B)} = r_e/(1 \pm B/B_e)$ of the ring electron in such an external magnetic field B means a proportional change $M_{e(B)} = M_e/(1 \pm B/B_e)$ of the electron magnetic moment, while its spin $L_{e(B)} = L_e = m_0 c r_e$ has to remain unchanged because of the fundamental law of angular momentum conservation. Or, such a spin conservation is possible only if in an external magnetic field B the electron has also two different rest masses $m_{0(B)} = m_0(1 \pm B/B_e)$, because only so its spin remains the same, $L_{e(B)} = m_0(1 \pm B/B_e) c r_e/(1 \pm B/B_e) = m_0 c = L_e$. But two different rest masses $m_{0(B)} = m_0(1 \pm B/B_e)$ of the same electron in the two possible orientations of its moments $\bar{M}_e$ and $\bar{L}_e$ in the external magnetic field $\bar{B}$ means inherently two different rest energies of the particle, $E_{0(B)} = E_0(1 \pm B/B_e)$, separated by the energy gap $\Delta E_0 = E_0 \cdot 2B/B_e$, or

$$\Delta E_0 = 2M_e B,$$

if we take into account that $B_e = m_0 c/e r_e$, $E_0 = m_0 c^2/2$ and $M_e = e c r_e/2$. And because the energy gap $\Delta E_0 = 2M_e B$ does not depend on velocity, it is valid not only for the electrons at relative rest, but also for all electrons in motion with velocities $v \ll c$, either free, as it has been directly noticed in "geonium" atom imagined by Ekström and Wineland, or bound in atoms, where their energy levels exhibit indeed both fine and hyperfine splits $\Delta E = 2M_e B$ in any intra-atomic or external magnetic field B.

Of course, this general experimental confirmation of the splitting formula $\Delta E = 2M_e B$ validates implicitly all equations previously deduced for the ring electrons, including not only that borrowed from the macroscopic accumulating rings $r_e = m_0 c/eB_e$, but also the resulted magnetic induction felt by magnetic preons on their subquantum orbit in the electron at rest, $B_e = 4.409 \cdot 10^9$ T.

As one can see, if the mass m_e of the ring electron raises no issue, since it is every time the sum of equal masses m_p of its constituent preons, $m_e = nm_p$, where n is a very big integer number, its attached quantum of electricity e remains a very intricate question: conforming to the basic laws of electromagnetism only this elementary electric charge attached to the electron can be the source of its dipole magnetic field, but, on the other hand, such a magnetic field can be generated only by a rotating circular electric charge, evenly spread on its perimeter. Or, as the elementary electric charge has been from the very outset a pointlike indivisible entity, its partition among all the preons of a ring electron, whose number can vary between very large limits, is something quite unimaginable. More, a circular quantum of electricity would have an electric field with axial symmetry, while in electrostatic theory this field has a spherical symmetry, the only one possible for a pointlike charge.

For all that, despite these apparently insurmountable difficulties for accepting an electricity quantum of circular form, in fact such an infringement of the so old electrostatic theory has been already assumed for a long time, as the only possible way for explaining the dipole magnetic field of the electrically neutral neutron. Indeed, all experimental research showed that the neutron is made of a small rotating hard core electrically positive, which contains almost the whole mass of the neutron, surrounded by a rotating coaxial circular mantle electrically negative, an aggregate whose resultant magnetic moment is that of the neutron. Another explanation for the electrically neutral but yet magnetic neutron simply cannot exist.

Or, by considering the known reactions of the neutron synthesis and decay, $n^0 \leftrightarrows p^+ + e^-$, as well as that of the neutron-antiproton annihilation, $n^0 + p^- \rightarrow \gamma + e^-$, it results clearly that the outer rotating ring of the neutron carries exactly one negative quantum e^- of electricity. And this accepted negative circular quantum of electricity inside the neutron proves the real physical possibility of a negative circular quantum of electricity carried by the free ring electron itself, hence the physical viability of this classical model of electron, in spite of its unavoidable non-conformity with that ancient concept of the pointlike electricity quantum, imagined in the prehistory of physics and preserved until now with no critical reexamination.

Ring electrons in extremely strong magnetic fields

Two decades after discovering the muon, Salam was qualifying it as the most mysterious elementary particle in physics, because there is no plausible explanation for its existence.

Mukhin[31] details this mystery in quantum physics of elementary particles:

"Concurrently with the accumulation of experimental data concerning the properties of the muon, it was outlined ever more clearly its extraordinary analogy with the electron. Indeed, the muons and electrons have identical spin ($s = 1/2$), and the same barion ($B = 0$) and electric ($z = \pm 1/2$) charges. Either of them takes part in the weak interaction with all its particularities (small section, violating the law of parity conservation). Neither of them participates in strong interactions. Either of them participates in a similar manner in electromagnetic interaction: for example, the μ^--meson, just like the electrons, can enter the composition of the atom, by creating the µ-mesoatoms. The energetic transitions of the μ^--meson in µ-mesoatoms are accompanied by electromagnetic radiation.

In a word, it is created the impression that the difference between the μ^--meson and the electron does not manifest itself otherwise than in values of their masses ($m_\mu \cong 207 m_e$), so that the muon is also called "heavy electron".

Enigma of muon mass is one of the most difficult problems in particle physics. Frankly speaking, here we have not one, but two questions:

1. Why, against the full identity between the properties of the muon and the electron, are their masses so different?

2. Why the muon mass is so big?

Let us clarify the second question. According to actual theoretical ideas, the magnitude order of mass is determined by the intensity of the strongest interaction the given particle partakes of. For the particles that participate in the strong interaction (nucleons, π-mesons, K mesons, hyperons) it is of hundreds of MeV/c^2; for the ones which participate in the electromagnetic interaction (and not in the strong one) it is of 1 MeV/c^2. Just this is the magnitude order of mass that the electron has. It would seem that the same mass should belong to the muon. But nature thought differently. The mass of muon is for unknown reasons 207 times greater than the mass of electron."

When Mukhin was writing the above, another essential difference between muon and electron was already known: their different magnetic moments, found by magnetic resonance to be exactly inversely proportional to their rest masses, $M_\mu/M_e = m_e/m_\mu$.

In fine, a last essential difference between muon and electron is their lifetime, because $\mu^\pm$-mesons change in flight into electrons $e^\pm$ having the same velocity (very close to c), motion direction and longitudinal spin polarization, and neutrinos whose energy and mass are equal to the energy and mass difference between initial muons and final electrons.

Just this mixture of many similarities and some few but essential differences made many physicists to think that muons are in fact electrons in unstable "excited" states, and this hypothesis is obviously in agreement with previous equations $m_{0(B)}$, $E_{0(B)}$, $r_{e(B)}$ and $M_{e(B)}$ valid for the ring electrons inside an external magnetic field B.

Indeed, this mystery of the muons simply disappears if the electrons are infinitesimal circular currents propagating in a vacuum, and therefore in an external magnetic field B their radius varies in accordance with the already deduced formula $r_{e(B)} = r_e/(1 \pm B/B_e)$, where $B_e = 4.409 \cdot 10^9$ T. For instance, when a spinning ring electron having a very high velocity $v \approx c$ and mass $m = m_0/\sqrt{(1-v^2/c^2)} > 207 m_0$ is stopped to a velocity $v \ll c$ in a space with huge magnetic induction $B = 206\,B_e \approx 9 \cdot 10^{11}$ T, wherein $\bar{B}_e \uparrow\uparrow \bar{B}$, the consequences are the following:

(1) Its radius decreases instantaneously from $r_e = 3.866 \cdot 10^{-13}$ m to $r_{e(B)} = r_e/207 = 1.868 \cdot 10^{-15}$ m ;

(2) Simultaneously its magnetic moment also decreases to $M_{e(B)} = M_e/207 = 4.409 \cdot 10^{-26}$ A·m² ;

(3) Its rest mass becomes $m_{0(B)} = 207 m_0$ by radiating the mass difference $\Delta m_{rad} = m - 207 m_0$ in order to preserve its immutable spin, the same in any external magnetic field, $L_{e(B)} = m_{0(B)} c r_{e(B)} = L_e = m_0 c r_e$.

Now it is easy to notice that the mass $m_{0(B)}$, radius $r_{e(B)}$, magnetic moment $M_{e(B)}$ and spin $L_{e(B)}$ of an electron at rest in a huge magnetic field $B \approx 9 \cdot 10^{11}$ T are equal to those of mysterious muon, in order

$$m_\mu = 207 m_0 = 4.400 \cdot 10^{-31} \text{ kg},$$
$$r_\mu = r_e/207 = 1.868 \cdot 10^{-15} \text{ m},$$
$$M_\mu = M_e/207 = 4.486 \cdot 10^{-26} \text{ A·m}^2,$$
$$L_\mu = L_e = 1.055 \cdot 10^{-34} \text{ J·s}.$$

Through this simple mechanism of contraction in a huge magnetic field, the light ring electron with high enough mass change into a ring muon stable in a huge magnetic field $B \approx 9 \cdot 10^{11}$ T, but unstable in a space destitute of high magnetic induction, where by dilation and radiating the mass

$\Delta m = 207m_0 - m_0$ it becomes again a light ring electron with rest mass m_0, radius r_e, magnetic moment M_e and spin L_e.

Similarly, when a ring electron having a very high velocity $v \approx c$ and mass $m = m_0/\sqrt{(1-v^2/c^2)} > 283m_0$ is stopped to a small velocity $v \ll c$ in a space with a higher magnetic induction $B = 282\, B_e \approx 1.25 \cdot 10^{12}$ T wherein $\bar{B}_e \uparrow\uparrow \bar{B}$, it becomes a charged pion at relative rest, whose main physical properties are in order

$$m_{\pi^\pm} = 283\, m_0 = 3.219 \cdot 10^{-31} \text{ kg},$$
$$r_{\pi^\pm} = r_e/283 = 1.366 \cdot 10^{-15} \text{ m},$$
$$M_{\pi^\pm} = M_e/283 = 3.281 \cdot 10^{-26} \text{ A} \cdot \text{m}^2,$$
$$L_{\pi^\pm} = L_e = 1.055 \cdot 10^{-34} \text{ J} \cdot \text{s}.$$

Of course, if this charged pion leaves its domain of stability, finally it becomes again an ordinary electron, but in this case after two distinct stages of dilation, $\pi^\pm \to \mu^\pm \to e^\pm$.

As seen, these reversible transformations $e^\pm \rightleftarrows \mu^\pm \rightleftarrows \pi^\pm$ in the external magnetic field explains why "the electron and μ meson behave like two states of one particle", and confirms the old supposition "they differ by a hitherto unknown internal parameter, presumably connected with their mass difference."[59]

An essential question is where the electrons strongly accelerated in electromagnetic field reach huge magnetic induction about 10^{12} T and even more: well, this always happens for a very short time when they collide in full with other magnetic particles, as slow nucleons and electrons, or, more concretely, whenever the very rapid and heavy electrons are scattered from slower nucleons and electrons. As known, for many decades the pointlike electron is upheld by the collision parameters noticed for the high-energy electrons scattered from slow nucleons or electrons, which lead to very small dimensions of the incident electrons, measured in attometer units (10^{-18} m). Or, such small radii of the electron utterly disagree with the known magnetic moment $M_e = 9.285 \cdot 10^{-24}$ A·m² of this particle, which requires a ring electron with radius of about 10^{-13} m magnitude order.

Unfortunately, none of those who invoke this real disagreement has remembered the old "flexible" ring electron sustained by Compton, whose

---

[59] M. Goldhaber, *Doubling of Fermions?*, Phys. Rev. Lett. **1**, (1958) 467.

dimensional dependence on the external magnetic field explains logically these experimental data: just because of its radius depending at each moment on the external magnetic field, the ring electron can in truth have extremely small radii at the moments of their collision with other magnetic particles (nucleons, electrons, etc.), in the close proximity of which the local magnetic induction could indeed reach huge values $\geq 10^{14}$ T. Evidently such advanced contractions of the incident fast electrons appears just at the moments of their closest approaching to the target particles, the only determinant for the colliding parameters noticed in these scattering experiments, but after coming again in a space devoid of strong magnetic induction the electrons revert *instantaneously* to their normal radius $r_e \approx 10^{-13}$ m. Indeed, if we take into account the estimated radius about $3 \cdot 10^{-16}$ m of the positively charged core of the neutrons[60], which is very near to that estimated now for the free proton, and consider it as being the minimum distance d_{min} betwen the incident fast electrons and the target nucleons at the moments of their frontal collision, then magnetic induction experienced by the incident high-energy electrons at the moments of their closest approaching to the positive core of the encountered nucleons can really be estimated to be $B_{max} \approx 2 \cdot 10^{-7} M_p/(d_{min})^3 \approx 10^{14}$ T, where $M_p = 1.41 \cdot 10^{-26}$ A·m² is the proton magnetic moment. Of course, only the electrons with very high energy can reach this maximal magnetic induction, otherwise the scattering experiments observe a general inverse proportionality between the velocity of the incident electrons and their size deduced from the colliding parameters, evidently because an initial higher momentum of the incident electrons allows their more advanced closeness to the target particles, where magnetic induction generated by the latter has higher values.

In addition, in such sudden collisions it has to be taken into account not only the advanced contraction of the electrons, but also that of the nucleons themselves, much weaker but not entirely non-existent at extremely high magnetic induction. Also, a non-spherical shape of the nucleons reduces significantly the minimum distance d_{min} and increase correspondingly the maximum magnetic induction B_{max} determinant for the size of the strongly contracted ring electrons bound by the protons as ring mesons within the neutrons.

---

[60] J-L. Basdevant, J. Rich and M. Spiro, *Fundamentals in Nuclear Physics*, Springer, 2005.

Anyway, even so there is an evident agreement between the magnitude order of magnetic induction necessary for shrinking the ring electrons to dimensions below 10^{-17} m and that maximally reachable for the heavy and rapid electrons during their collisions with slow nucleons, both of them around 10^{14} T.

As for the scattering experiments involving two opposite fascicles of electrons highly accelerated in magnetic field, the smaller minimum dimensions about 10^{-19} m calculated from the colliding parameters experimentally noticed are also explainable, because besides the much higher magnetic moment of the target electrons, in comparison with those of the target nucleons, in this case both particles involved in a frontal collision are very strongly contracted.

More, this contraction of both ring electrons in collision allows a much more profound approaching of them in comparison with the case of the electrons scattered from the much less contractible nucleons, and finally an increased magnetic induction felt by the incident electrons at the moment of their collision, even if at that moment their magnetic moments are about 10^5 times diminished.

Magnetic nucleons

When fast electrons are scattered from slow nucleons, only the electrons exhibit dimensions much below their *normal* size $r_e = 3.866 \cdot 10^{-13}$ m, and this advanced contraction is possible just because they are circular electric currents *hollow* inside, while the nucleons can change much less their dimensions because the subquantum circular currents originating their dipole magnetic fields with axial symmetry surround much smaller and heavier granular entities, compressed and stiffened by the strong interactions acting between their subcomponents called partons by Feynman or quarks by Gell-Mann.

Therefore, if the proton really consists of a small, heavy and neutral core surrounded by the stationary circular current responsible for the dipole magnetic field of the particle, then this outward circular current assesses not only the spin magnetic moment of the whole composite proton, $M_p = ecr_p/2 = 1.410 \cdot 10^{-26}$ A·m², but also its radius

$$r_p = 2M_p/ec = 0.59 \cdot 10^{-15} \text{ m}.$$

And this subquantum circular current in the proton can only be a ring positive meson stable in the huge magnetic field around the compact core of this composite particle.

First experimental data regarding the size of the proton date from the 1950's, when the resulted dimensions were situated approximately between $0.7 \ldots 0.8 \cdot 10^{-15}$ m [61], but in the next decades the proton size decreased to about $0.5 \cdot 10^{-15}$ m [31]. Anyhow, the proton radius $r_p = 0.59 \cdot 10^{-15}$ m calculated from the known magnetic moment of the particle by general formula $r = 2M/ec$ is in good agreement with all experimental data in the field, irrespective of the used method.

As for the inner structure of the neutral but yet magnetic neutron, since several reactions, for example

$n \to p^+ + e^- + \nu_e$ (neutron decay in free state or in β-active nuclei),

$\pi^- + d^+ \to 2n + \nu_\pi + \gamma$ (charged pion capture in deuteron[62]),

$n + p^- \to \gamma + e^-$ (neutron-antiproton annihilation[63]),

point out its composite structure proton-negative meson, the latter becoming by dilation a ring electron in free state, the contoured picture of the neutron is similar to that already accepted in quantum physics, but by replacing the evasive term "circular cloud of electricity" (used to avoid the more exact term "circular quantum of electricity", much too undesirable for evident reasons) with charged ring meson.

Therefore, the neutron has to be seen as a plane structure, made of an exterior negative ring meson and an inner proton, in turn made of a central small, hard and neutral core surrounded by a positive ring meson, or, in other words, the neutron is made of a small, hard and neutral core surrounded by two concentric ring mesons, the inner positively charged, the external negatively. Evidently the radius of the outward negative ring meson is that of the whole neutron, and for calculating it we have two variants:

---

[61] E. E. Chambers and R. Hofstadter, *Structure of the Proton*, Phys.Rev. **103**, (1956) 1454.

R. Hofstadter, *Nuclear and Nucleon Scattering of High-Energy Electrons*, An. Rev. Nucl. Sci. **7**, (1957) 231.

[62] W. Chinowsky and J. Steinberger, *Absorption of Negative Pions In Deuterium: Parity of the Pion*, Phys. Rev. 95,(1954) 1561.

[63] E. Segrè, *Antinucleons*, Annu. Rev. Nucl. Part. Sci. **8**, (1958) 127.

(1) If the inner ring proton in the neutron has a radius and magnetic moment not much different from those of the free proton, $r_p = 0.59 \cdot 10^{-15}$ m and $M_p = 1.410 \cdot 10^{-26}$ A·m², then the ring electron fastened by the proton in a coaxial and coplanar coupling, and strongly contracted to meson state, has to have an magnetic moment about $M_m \approx -(0.97 + 1.41) \cdot 10^{-26}$ A·m² $\approx -2.38 \cdot 10^{-26}$ A·m², because only so the composite neutron can have in aggregate its known magnetic moment $M_n = -9.662 \cdot 10^{-27}$ A·m². If so, the radius of the exterior negative ring meson in the neutron is

$$r_m = 2M_m/ec = 0.99 \cdot 10^{-15} \text{ m},$$

which is evidently that one of the whole neutron;

(2) If the radius of the inner positive core of the neutron can be experimentally estimated to about $r_p \approx 3 \cdot 10^{-16}$ m [60], which means that the central proton inside the neutron is strongly contracted as against its normal radius $r_p = 0.59 \cdot 10^{-15}$ m, and consequently has a much smaller magnetic moment $M_p \approx 0.72 \cdot 10^{-26}$ A·m², then the peripheral negative ring meson inside the neutron has to have its own magnetic moment of about $M_m \approx -(0.97 + 0.72) \cdot 10^{-26}$ A·m² $\approx -1.69 \cdot 10^{-26}$ A·m², and therefore a radius

$$r_m = 2M_m/ec \approx 0.70 \cdot 10^{-15} \text{ m},$$

which in this variant becomes that of the whole neutron.

Both calculated radii can be considered to agree to those experimentally deduced in the course of time by different methods, especially the second one $r_m \approx 0.70 \cdot 10^{-15}$ m, very near to the Hofstadter's results. Also, in both variants the neutron is more or less larger than the proton, which is logical for a more complex structure, even if the proton encapsulated inside the neutron is in truth significantly contracted (of about 50 %) as compared to the free proton. And indeed, a larger size of the neutron as against that of the proton was experimentally found[64].

The plane structure previously deduced for the two composite nucleons decreases significantly the minimum distance at which they can approach coaxially each other, and this means a rapid increase of the energy of their attractive magnetic interaction, which observes a law in $1/r^3$ and

---

[64] W. K. H. Panofsky and E. A. Allton, *Form Factor of the Photopion Matrix Element at Resonance*, Phys. Rev **110**, (1958) 1155.

W. Hirt, *Neutron-proton radius difference and pion production by protons on nuclei*, Nucl. Phys. **B9**, (1969) 447.

can so reach values even to some MeV, hence at the level of the known binding energies of the nuclear nucleons, which disaffirms clearly the old conviction that the nuclear forces acting between nuclear nucleons are too strong to be of magnetic nature.

The best quantitative example in this regard is just the deuteron, made of one neutron with magnetic moment $M_n = -0.967 \cdot 10^{-26}$ A·m² and one proton with magnetic moment $M_p = 1.410 \cdot 10^{-26}$ A·m², whose dipole magnetic fields have coaxial symmetry axes, which is proved by the fact that the deuteron magnetic moment $M_d = 0.433 \cdot 10^{-27}$ A·m² is almost equal to the arithmetic sum of the two magnetic moments of its constituent proton and neutron, $M_p + M_n = 0.443 \cdot 10^{-27}$ A·m². However, before verifying if this magnetic coaxial coupling can justify the binding energy corresponding to the mass defect of deuteron $\Delta m = (m_n + m_p) - m_d = 0.00240$ u, where $m_n = 1.0086$ u, $m_p = 1.00728$ u and $m_d = 2.01355$ u are the known masses of neutron, proton and deuteron, we must remember that this binding energy has to be calculated by the classical formula $E = \Delta mc^2/2$ after halving the Planck's constant to $h = 3.313 \cdot 10^{-34}$ J·s, and consequently the real binding energy of the deuteron is equal to $1.78 \cdot 10^{-13}$ J (or 1.11 MeV). Or, if this binding energy is equal to that of the attractive magnetic interaction between the two coaxial dipole magnetic fields existing in deuteron, $E = 2 \cdot 10^{-7} M_p M_n/d^3$, where d is the distance between the parallel symmetry planes Oxy of these fields, it results an entirely plausible distance proton-neutron $d = 0.535 \cdot 10^{-15}$ m inside the deuteron, some smaller than the charge radii of the two nucleons in their free state, but quite possible if the two nucleons in deuteron have their magnetic moments coaxially coupled. Moreover, this distance neutron-proton inside the deuteron tallies with the action radius of the strongly repulsive forces acting between nuclear nucle ons at distances $< 5 \cdot 10^{-16}$ m from each other, whose existence was found in scattering experiments $(n-p)$ and $(p-p)$ at energies over 100 MeV [31], but whose origin is still unknown. As normal, the distance between the two coaxial components of the deuteron has to be about equal to the action radius of these strong repulsive forces between two nucleons[65].

Although the known decay of the free neutron demonstrates the weak enough connexion between its outward negative ring meson and the central

---

[65] Very interesting, these very strong repulsive forces with small radius of action do not act yet between neutron and antiproton, which annihilate together!

positive proton, from which this outward ring meson can separate at a given moment, inside the deuteron such a decay of its constituent neutron does not occur. And indeed, this stability of the deuteron becomes easy to explain by considering it as being made of three circular and coaxial magnetic subcomponents [proton – negative meson – proton], with a central negative meson with larger radius continuously interchanged by the two marginal protons with smaller radius, whose identity change alternatively from neutron into proton and vice versa, $n \rightleftarrows p$.

More, this tripartite coaxial structure [positive ring proton – negative ring meson – positive ring proton] of the deuteron, whose binder is just the central negative meson brought by initial neutron and then equally shared by the two marginal protons within deuteron, between of which this central negative ring meson has a continuous motion back and forth, can explain intuitively not only why the neutron coupled with a proton does not decay like the neutron alone, but also other two very peculiar questions resulted from experimental research:

(1) Why the attractive force between proton and neutron appears as an *exchange* force;

(2) How is possible, within certain limits, a direct proportionality between the distance proton-neutron and the strength of their attraction force, a dependence almost unique in all known interactions between elementary particles, but which could be due to the already several times mentioned direct proportionality between the strength of the dipole magnetic field generated by a ring particle and its radius, because thus the dipole magnetic field of the negative meson in the neutron is the more stronger as the distance proton-neutron is larger, and consequently its contraction is smaller.

On the other hand, as deuterons remain magnetic dipoles with determinate magnetic moment, they can also bind each to other with coaxial and antiparallel magnetic moments in order to form α-particles with null magnetic moment. And not at all accidentally two non-magnetic α-particles cannot bind together any longer in order to form a stable nucleus $^{8}_{4}Be$ [66], which needs a supplementary magnetic neutron.

---

[66] In fact, the unstable isotope $^{8}_{4}Be$ has yet a total binding energy of about 0.11 MeV, but this is entirely negligible as against the mean binding energy between two nuclear nucleons, of about 60 times higher. And this still exists only because of the residual magnetic interactions between the marginal nucleons of the two α-particles.

More, as these non-magnetic α-particles are present in all the more complex nuclei, inside these nuclei only the nucleons or deuterons supplementary to these quasi-inert (in aggregate) α-particles show to generate nuclear forces by which they can interact in different nuclear reactions with other active nuclear entities, always magnetic. Long time contested, although the first clear experimental proofs for the real existence in nuclei of the α-particles were found since long ago[67], just because the nuclear inactivity of these non-magnetic nuclear entities proves the real nature of the nuclear forces, eventually the presence of α-particles in all the heavier nuclei became a certitude recognized as such.

In addition, this magnetic nature of the nuclear forces acting between free or nuclear nucleons is manifestly reinforced by some known features of the nuclear forces, but typical for magnetic interactions:

(1) Their small radius of action, because the forces between magnetic dipoles observe a law in $1/r^4$, with a very rapid decrease of their intensity with distance;

(2) Their dependence on the mutual orientations of nuclear magnetic moments;

(3) The presence of the spin-orbit interactions that proves the dependence of the nuclear forces on the orbital velocities of the nucleons.

Still the clearest argument for magnetic nature of nuclear forces remains evidently the extremely weak nuclear interaction between the non-magnetic α-particles. By their exactly null magnetic moment, these particles prove to consist in two magnetic deuterons paired coaxially with antiparallel magnetic moments. Or, we see that in the absence of an own magnetic moment proper to an own dipole magnetic field the α-particles are practically unable to bind each to other in order to form a stable nucleus $^{8}_{4}Be$, while their constituent neutrons, protons and deuterons, all of them having their magnetic moment, interact very strongly through nuclear forces.

Resuming, the structural models previously sketched, namely proton made of a small, hard, heavy, neutral and spinless core surrounded by a negative spinning ring meson, and neutron made of a proton surrounded in

---

[67] G. Igo, L. F. Hansen, and T. J. Gooding, *Evidence for "Alpha-Particle" Clusters in Several Nuclei from (α, 2α) Reaction at 0.91 BeV*, Phys. Rev. **131**, (1963) 337.

H. Gauvin, M. Lefort and X. Tarrago, *Mise en évidence de structures alpha dans les noyaux lourds*, Phys. Lett. **4**, (1963) 215.

plus by a coplanar negative spinning ring meson, prove to be able not only to explain the main properties of the two nucleons, from their size to their magnetic moment, but also to demonstrate convincingly the magnetic nature of the strong nuclear forces with very small action radius by which they interact between them inside nuclei.

Ring electrons accelerated in electromagnetic field

In the previous chapter it was brought forward the manifest incapacity of special relativity to ensure a simultaneous conservation of energy and momentum when the free or atomic electrons absorb or radiate photons, although the later are the acknowledged carriers of electromagnetic interaction at a distance (the relativistic interpretation of photoelectric effect cannot be taken seriously). Of course, such a failure would also be unacceptable for the ring electron model, and in this case everything has to be in compliance with the fundamental laws of conservation in classical physics.

Therefore, when a ring electron initially at rest absorbs a photon with velocity c and mass m_γ, and the whole energy of this absorbed photon is converted into kinetic energy by the electron, the final velocity v of the accelerated electron is determined by the equation of energy conservation $m_\gamma c^2/2 = (m_0 + m_\gamma)v^2/2$, whence it results $(m_0 + m_\gamma) = m_0/(1 - v^2/c^2)$. In these conditions the inner energy E_i of the ring electron has to remain unchanged, but this is possible only if the inner peripheral velocity v_i of the accelerated electron decreases simultaneously from initial value c at rest to a value determined by its independent equation of energy conservation, $E_i = m_0 c^2/2 = (m_0 + m_\gamma)v_i^2/2$, wherefrom it results $v_i = c\sqrt{(1 - v^2/c^2)}$.

However, if an elementary act of photon absorption would end so, the accelerated ring electron would have a spin larger than that at rest, $L_e = [m_0/(1 - v^2/c^2)]c\sqrt{(1 - v^2/c^2)}\, r_e = m_0 c r_e / \sqrt{(1 - v^2/c^2)} > L_e = m_0 c r_e$, what would infringe the universal law of angular momentum conservation. Accordingly, the ring electron has in turn to radiate a photon during its electromagnetic acceleration, until its mass decreases exactly at the level $m = m_0/\sqrt{(1 - v^2/c^2)}$ necessary for keeping the immutable spin of the electron in conformity with the universal law of angular momentum conservation, $L_e = m v_i r_e = m_0 c r_e$.

In consequence any ring electron accelerated in electromagnetic field by photon absorption has a final mass

$$m = m_0/\sqrt{(1 - v^2/c^2)},$$

as a requirement of the same universal law of angular momentum conservation, and this conclusion is valid irrespective of the initial velocity of the accelerated ring electron, if this electron is or not at rest before its electromagnetic acceleration.

For the same reasons the mass equation $m = m_0/\sqrt{(1 - v^2/c^2)}$ is valid also for the charged pions, muons and β-electrons, all of them initially electrons strongly accelerated in electromagnetic fields.

As one can see, the ring electron accelerated in electromagnetic field has a mass dependence on velocity similar to that of the relativistic pointlike electron, but the reasons are quite different. While in special relativity this mass dependence on velocity is an universal relativistic effect, valid unconditionally for all elementary particles irrespective of their origin and history, the ring electron model predicts this equation only for elementary particles accelerated in electromagnetic field (electrons, nucleons, mesons, etc.). Or, the mass defect of nuclear nucleons and atomic electrons invalidates clearly the relativistic comprehension of mass equation $m = m_0/\sqrt{(1 - v^2/c^2)}$.

Differently from the pointlike electron of quantum mechanics, whose rest mass, magnetic moment and angular momentum have intrinsic, immutable values, the classical ring electron has dimensions, rest mass and magnetic moment depending on the external magnetic induction, only its spin remains always the same due to the fundamental law of angular momentum conservation. And this spin invariance is responsible for changing the electrons into mesons, as well as for the photon radiation and the mass equation $m = m_0/\sqrt{(1 - v^2/c^2)}$ valid exclusively for elementary particles accelerated through photon absorption.

Still the most special characteristic of the pointlike electron is undoubtedly its dualistic behavior in substance: when this electron passes through an atomic lattice, it behaves like a corpuscle during its collision with an atomic electron, but like a wave if during its travel within the highly ordered lattices such collisions do not occur (excepting the perfect reflections on some imaginary, non-existent interstitial planes, the basic hypothesis of the theory of Bragg diffraction).

As this wave-corpuscle duality is inconceivable for the classical ring electron, some critical remarks on this concept are necessary.

Wavicle electron, reality or fantasy much too easy accepted?

Soon after the Planck's quantization of the energy radiated by a blackbody, the second step in this direction was a partial return to the Newton's corpuscular vision on light, but without denying its wave nature. Einstein was one of the first adherents to this dualism of light:

"It seems to me that the observations associated with blackbody radiation, fluorescence, the production of cathode rays by ultraviolet light, and other related phenomena connected with the emission or transformation of light are more readily understood if one assumes that the energy of light is discontinuously distributed in space. In accordance with the assumption to be considered here, the energy of a light ray spreading out from a point source is not continuously distributed over an increasing space but consists of a finite number of energy quanta which are localized at points in space, which move without dividing, and which can only be produced and absorbed as complete units."[68]

"Newton's corpuscular theory is reanimated again, although it proved to be completely unsound in the domain of geometrical properties of light. We have now, therefore, two theories of light; both are necessary, and both, we have to admit this, exist without any logical connection, despite the twenty years of tremendous efforts of theoretical physicists."[69]

Well, in 1924 De Broglie[70] made the last move on this way and proposed a generalized dualism wave-particle, conforming to which any elementary particle is not only an infinitesimal corpuscle, with mass, momentum and kinetic energy, but also a wave whose wavelength is $\lambda = h/mv$, where m and v are in order the mass and the velocity of the particle. And when in 1928 G. P. Thomson[71] published his final experimental results on the

---

[68] A. Einstein, *Über einen die Erzeugung und Verwandlung des Lichtes betreffenden heuristischen Gesichtspunkt*, Ann. Physik **17**, (1905) 132.

[69] A. Einstein, *Das Comptonsche Experiment*, Berliner Tageblatt (20 April 1924).

[70] L. de Broglie, *Recherches sur la théorie des quanta*, Thesis (Paris), 1924 and Ann. Physique **3**, (1925) 22.

[71] G. P. Thomson, *Experiments on the Diffraction of Cathode Rays*, Proc. Roy. Soc. London. **117**, (1928) 600.

electron diffraction through *Al* and *Au* foils with thickness below 10^{-8} m, when alternatively light and dark concentric rings appeared on the photographic plates behind the foils, resembling in many respects to those previously noticed in similar experiments using X-rays, the effect was so lightning that de Broglie was awarded the Nobel Prize next year!

What is amazing in this story is undoubtedly the easiness in acknowledging with no precautionary measure an idea utterly disconcerting for understanding the matter. And first ignored alternative has been the photoelectric effect, so much experimentally studied in the previous years.

Indeed, then it was already known that the low-energy electrons eject photoelectrons from a metal lattice on directions perpendicular to their motion direction, but when their energy increases the emission angles decreases progressively. At least for this reason the photoelectrons certainly emitted from a thin metal foil staffed by energetic electrons ought to have been considered as possibly responsible for the concentric rings appeared on the photographic plates, but inexplicably no minimal verification of this logical version existed.

This complete ignoring of the photoelectric effect for interpreting the images of electronic diffraction through metal thin foils is downright hard to understand, because the positive charge of irradiated metals was already known at those days, including the fact that an intense solar radiation can electrify thin plates of gold, silver or copper strong enough to electrocute persons in direct ground connection.

And even if the electronograms are not directly drawn by the secondary photoelectrons, the incontestable presence of a lot of differently polarized Me^{n+} atoms at the emergence end of the interstitial path of the incident electrons could obviously deviate the trajectory of the latter at the moments when they leave the thin metal foils, and each deviation depends obviously on the attraction power exerted on the respective electron by the ionized atom Me^{n+} near to which it passes. In other words, how many kinds of Me^{n+}-ions exist at the emergence end of the interstitial channels crossed by the incident particles, so many different angles of deviation they should cause, and evidently so many diffraction rings should appear on the final photographic plates.

How necessary such a research would have been results from a simple comparison between the growth rates of the concentric diffraction rings in electronograms obtained with Ag, Au, Cu or Al thin foils, and the square roots of the successive ionization potentials of these metals (in the data below the diffraction rings diameters d_1, d_2, ... d_n and the square roots of ionization potentials $\sqrt{I_1}$, $\sqrt{I_2}$, ... $\sqrt{I_n}$ are given in relative units, therefore $d_1 = 1$ and $\sqrt{I_1} = 1$). Of course, I_1 corresponds to neutral atoms Me^0, I_2 to Me^+-ions, I_3 to Me^{2+}-ions, and so on.

$$Ag - d_n - 1 / 1.64 / 2.03 / 2.55 / 3.03 / 3.51$$
$$\sqrt{I_n} - 1 / 1.68 / 2.18 / 2.42 / 3.03 / 3.42$$
$$Au - d_n - 1 / 1.55 / 1.86 / - / 2.42 / 2.96$$
$$\sqrt{I_n} - 1 / 1.49 / 1.82 / 2.17 / 2.50 / 2.82$$
$$Cu - d_n - 1 / 1.66 / 1.96 / 2.76 / 3.30 / 3.68 / 4.29$$
$$\sqrt{I_n} - 1 / 1.62 / 1.95 / 2.76 / 3.20 / 3.77 / 4.37$$
$$Al - d_n - 1 / 1.76 / 2.20$$
$$\sqrt{I_n} - 1 / 1.77 / 2.18$$

Or, in spite of the lower accuracy of these measurements directly on the electronograms found in the most accessible literature (books, treatises, articles)[72], in the table above it is very clear the pregnant resemblance between the growth rates of the diffraction rings and those of the square roots of the ionization potentials, and these analogous rises prove convincingly that the electron diffraction through thin foils is closely related, one way or other, to the ionized atoms Me^{n+} existing certainly in any metal lattice staffed by any kind of particles of high energy.

Even the number of concentric rings in electronograms generally corresponds to that of the electrons placed on the outer electronic layers of the atoms in the used metals, the only removable from these atoms just because their binding energies are smaller than the energy of the incident particles.

Thus, while all electronograms through heavy metals (Ag, Au, Cu or Pt) consist in numerous diffraction rings, those through Al foils exhibit only three, therefore exactly the number of electrons existing on the valence layer of Al atoms. More, when these electronograms are relatively much enlarged, the first ring splits into two very close distinct rings, as a kind of fine splitting of this smallest ring, which has a clear correspondence to the

---

[72] For comparison, in his electronograms much less clear and with fewer concentric rings, G. P. Thomson measured 1/1.4/2.08 for Al and 1/1.72/2.03/2.82/3.360/- /4.44 for Au.

different binding energies of the three valence electrons in Al atoms, namely a little higher for the two magnetically paired electrons in the marginal electronic layer of the atom, as against the third unpaired electron. In other words, the fine splitting of the binding energy of the three valence electrons in Al atom is emphasized by the fine splitting of the smallest diffraction ring in Al electronograms.

But further on, in the absence of some experiments necessary for clarifying concretely how these ionized atoms Me^{n+} lead finally to the diffraction rings, now two possible mechanisms of corpuscular diffraction through very thin foils still can be had in view at first sight:

(1) The diffraction rings are shaped indeed by the incident particles differently deviated by the different Me^{n+}-ions localized around the emergence points of the incident particles;

(2) The diffraction rings are shaped in all cases by the secondary electrons ejected by the incident particles from ionized atoms Me^{n+} on directions dependent on the attraction power of the remained ions $Me^{(n+1)+}$.

As it was said above, the true variant can be assuredly established only by some appropriate experiments. And before all, the most necessary experiment would be now an electron diffraction through a thin metal foil connected to earth, or even to a weak source of free electrons, when a possible disappearance of the diffraction rings, or only their significant weakening in intensity, would definitely confirm the determining role of the Me^{n+}-ions in all kind of diffraction through thin foils (and not only).

Also, measuring the polarization degree of the electrons leaving a thin metal foil would certainly be telling from many viewpoints. As known, passing the unpolarized electrons through an atomic lattice extremely ordered is even a method used for obtaining spin polarized electrons. Or, such an polarization effect of the incident electrons initially unpolarized could lead to a new interpretation of the electronograms through mono- and polycrystalline foils (first known as Laue diffractograms) since any change of initial spin polarization of the incident particles means an angular momentum variation, of them, which can be compensated, conforming to the universal law of angular momentum conservation, only by inducing new orbital angular momenta of these particles, therefore by an assessed change of their motion direction.

Similarly, diffraction experiments with already polarized incident electrons passed through mono- and polycrystalline foils could be likewise significant for estimating the consequences of all changes in spin polarization experienced by the incident particles within atomic lattices with very high ordering degree of their structure, as respects their motion direction after leaving the thin metal foils.

As a matter of fact, many other experimental realities, as the interference fringes obtained by passing linear beams of electrons through fine electrical or magnetic devices[73], or even the electron microscopes, already proved that the wave behavior of electrons is in fact determined by their deviation in electric or magnetic fields. More, the electronograms through gases invalidate definitively de Broglie's dualistic theory, despite all ridiculous attempts to consider the free atoms or molecules in continuous and chaotic movement as "rudimentary" or "incipient" diffraction gratings. In exchange, these hazy electronograms through gases can still have a meaning as being consequences either of the photoelectric effect itself, or of the definite presence of some differently ionized atoms able to deflect under determinate angles the initial trajectory of the incident electrons electrons.

---

[73] G. Möllenstedt and H. Düker, *Beobachtungen und Messungen an Biprisma-Interferenzen mit Elektronenwellen*, Z. Phys. **145**, (1956) 377.

PHOTON RADIATION OF SPINNING RING ELECTRONS

Beam photon

As concluded formerly, the spinning ring particles radiate outside a part of their constituent preons under the form of photons, either during their electromagnetic acceleration by photon absorption, or when they are braked and consequently their inner linear velocity $v_i = c\sqrt{1 - v^2/c^2}$ increases as their velocity v decreases, because such photon emissions are the only possibilty of them for preserving unchanged their immutable spin $L_e = m_0 c r_e$. In short, any photon emission of a ring particle is induced by the universal law of angular momentum conservation.

The next step is to find how a preonic photon springs from a ring electron. This implies to identify not only a photon structure able to explain all the known properties of this elementary particle, some of them entirely peculiar, for instance its own frequency, but also a specific mechanism of its emission.

The simplest case is evidently that of a radiating ring electron at relative rest ($v \ll c$), whose constituent preons equally distanced from each other on their subquantum orbit have the same linear velocity $v_i \approx c$. This ring electron can emit a photon having a stable structure in time, at least in the void space, only if at a given moment a certain number of its constituent begin leaving their subquatum circumference at equal intervals of time and move inertially in the free space, all of them from the same point of their former subquantum circumference and with the same linear velocity $v_\gamma \approx c$. Therefore, a preonic photon radiated by a ring electron can only be a beam (or linear fascicle) of preons evenly spread on a straight line tangent to the subquantum orbit of the radiating ring electron, all preons in inertial motion along the straight line joining them, a line which conventionally can be called the axis of the beam photon.

More, we can consider the equal time intervals T at which the radiated preons leave the radiating ring particle as the constant emission period, the distance $\lambda = cT$ as the constant wavelength of the linear beam photon, and the quantity $\nu = 1/T$ as the constant emission frequency of this beam photon. And since all these quantities belong to wave physics, right from the start the strictly corpuscular model of beam photon suggests so the tempting possibility to replace the unreasonable wave-corpuscle dualism assigned to photons with the beam structure of these elementary particles, all the more so as the wave behavior of the linear beams of electrons, protons neutrons or atoms was noticed for long, for example their interference even when they propagate in free space, or their diffraction when they pass through thin metal foils, and so on.

As for the real case of the radiating ring electrons in uniform motion with linear velocity v, the above sketched mechanism of photon emission remains in principle the same, still some detectable changes appear:

(1) The velocity of the radiated preons is not always equal to c any longer, as long as their velocity $\bar{v}_\gamma$ is always equal to vectorial sum of the velocity $\bar{v}$ of the radiating ring electron and the inner linear velocity $\bar{v}_i$ of the preons on the subquantum orbit of this radiating ring electron at the moment of their detachment from the latter,

$$\bar{v}_\gamma = \bar{v} + \bar{v}_i ,$$

where $v_i = c\sqrt{1 - v^2/c^2}$, so that between their vector velocities $\bar{v}$ and $\bar{v}_i$ there is an angle depending on the spin polarization of the radiating ring electron.

(2) The motion direction of the linear beam photons radiated by moving ring particles does not coincide any more with their axis;

(3) An observer at rest measures a different frequency of the linear beam photon emitted by a moving ring particle, as against the frequency noticed for the linear beam photon emitted with the same emission period by the same ring particle when at rest, and this difference assessed by the motion of the radiating electron can be called "Doppler effect".

One notices easily to this mechanism of photon emission an aspect in disagreement with usual convictions, namely the absence of a mechanical recoil for the radiating electron, but actually this fact contravenes no experimental data, whereas such an emission recoil was never directly or indirectly revealed. Perhaps synchrotron radiation is the most conclusive example in this respect, because the electrons electromagnetically accelerated in synchrotron emit high-energy photons both inside the accelerating electromagnetic cavities, where they always radiate almost right forward with no effect on their linear velocities, and even after leaving these cavities, when they emit photons in all directions in their orbital plane, including forward (in their motion direction) and back (in the opposite direction of their motion), also with no change of their linear velocity or their strictly circular trajectory. Impossible when electrons with no inner structure radiate photons also with no inner structure, as in quantum relativistic physics, such photon emissions with no emission recoil for the radiating electrons is understandable yet if the spinning ring electrons radiate preons whose linear velocity $\bar{v}_\gamma$ after leaving the subquantum orbits of their emitters is the same with the total linear velocity ($\bar{v} + \bar{v}_i$) had by them on that former subquantum orbits at the moments of their detachment from the radiating ring electrons.

But the most important test for this preonic model of beam photon is undoubtedly its ability to justify the known proportionality $E_\gamma = h\nu$ between the photon energy and frequency, just because this entirely peculiar aspect is an essential property of the photons, the only elementary particles not only with an own frequency experimentally measurable, but much more, an own frequency directly proportional to their energy (which becomes thus a physical quantity easy to be determined).

Well, this is entirely possible if it is accepted a unique emission time τ_0 for all ring electrons whose velocity v is negligible compared to the velocity of light c, and consequently the same length $l_0 = \tau_0 c$ of all beam photons radiated by them. Indeed, in this case the number of preons in all beam photons depends only by their emission frequency, which means implicitly a direct proportionality $E_\gamma = h\nu$ between their energy E_γ (directly proportional with the number of their constituent preons) and their emission frequency ν.

And these unique values for spatial length l_0 and emission time τ_0 of all beam photons emitted by slow ring electrons can be accurately calculated based on some known experimental data, as long as this unique length l_0 is equal to the maximum wavelength λ_{max} of the simplest beam photon, made of only two preons, and this maximum wavelength corresponds to the smallest frequency $\nu_{min} = c/\lambda_{max}$ of all photons existing in nature, which in turn can be identified with that bottom limit frequency of the universal background of the microwaves (placed at the lower end of the electromagnetic radiation scale), $\nu_{min} = 4.08 \cdot 10^8$ Hz, experimentally found by Penzias and

Wilson[74] (it should be noticed that all radio frequencies below this inferior limit of the microwave range are in fact those of the alternative voltages in the mechanisms of antenna, and not at all those of the photons radiated by the conduction electrons successively accelerated in opposite directions in the spaces wherefrom the radio waves are emitted, photons whose frequency depends only on the amplitude of the electromotive forces acting on the radiating electrons).

Therefore, the length of all beam photons radiated by ring electrons with small velocities $v \ll c$ is

$$l_0 = \lambda_{max} = c/\nu_{min} = 0.735 \text{ m},$$

and their unique time of emission is

$$\tau_0 = l_0/c = 2.451 \cdot 10^{-9} \text{ s}.$$

Although at first sight a macroscopic size of an elementary particle seems to be something completely absurd, the standard length $l_0 = 0.735$ m found for all the beam photons emitted by slow ring electrons agrees pretty well with that so-called "wave train" (in fact an irrational wave with limited length about 1 m) attached in quantum physics to any pointlike photon in order to explain its undulatory behavior, for example some interference phenomena experimentally noticed:

"Through these arrangements, the wave is split into two waves packets which are later made to recombine after having traveled along different optical paths. Obviously, the above wave packets cannot interfere unless the difference between the two optical paths is smaller than spatial extension of the wave. In such phenomena, the spatial extension of the wave is clearly exhibited ...

Interference patterns corresponding to optical differences of the order of a meter have actually been observed. This confirms the conclusions which are drawing here on the spatial extension of the wave train, and on the ambiguity of the notion of decay time."[75]

---

[74] A. A. Penzias and R. W. Wilson, *A Measurement of Excess Antenna Temperature at 4080 Mc/s*, Astrophys. J. **142**, (1965) 419.

M. S. Longair and R. A. Sunyaev, *The universal electromagnetic background radiation*, Usp Fiz. Nauk. **105**, (1971) 41. (in Russian)

A. Webster, *The Cosmic Background Radiation*, Sc. Amer. **231**, (1974) 26.

[75] A. Messiah, *Quantum Mechanics*, Vol. I, North-Holland Publishing Company, Amsterdam, 1967.

Similar macroscopic lengths for the light quanta were found from the noticed width of spectral lines[76], or from the time necessary for atomic electrons in metal lattices to absorb the incident light quanta in photoelectric experiments[77].

On the other hand, the calculated emission time $\tau_0 = 2.451 \cdot 10^{-9}$ s can be connected with a long series of experimental data:

(1) The slow positrons have similar lifetimes in substance, for example in rare gases at very low temperatures[78], when the parasite influences due to the polar moments and the thermal photons are much diminished and consequently the measuring accuracy is the best: thus the mean lifetimes of $1.83 \cdot 10^{-9}$ s and $2.60 \cdot 10^{-9}$ s were experimentally estimated in liquid helium, $2.2 \cdot 10^{-9}$ s in solid krypton, and $2.7 \cdot 10^{-9}$ s in solid xenon;

(2) The lengths of time about some few nanoseconds of all electronic transitions $2p \to 1s$ in the H-like and alkaline atoms with only one electron on their peripheral layer, or of the upper states in the optical bands specific to molecular spectra, especially when they are measured by methods nearer to a direct timing, for example by delayed coincidence. These transitions are the least exposed to some perturbing factors that can greatly alter their real lifetime, and their review shows that the most experimental values of their lifetimes are expressed in nanoseconds;

By comparison, the time necessary for atomic electrons to jump from one orbit to other during an electronic transition is somewhere in the range $10^{-14} \ldots 10^{-15}$ s, a while absolutely negligible in the total length of time of an excited state in atom;

(3) The lengths of time equal to some few nanoseconds measured for the electron and ion bursts obtained either by electromagnetic acceleration or by electrical discharges in gases, as well for the light pulses appearing in the latter;

---

[76] G. P. Thomson, *Test of a Theory of Radiation*, Proc. Roy. Soc. A **104**, (1923) 115.

[77] E. O. Lawrence and J. W. Beams, *On the Nature of Light*, Proc. Nat. Acad. Sci. USA **13**, (1927) 207.

[78] D. A. L. Paul and R. L. Graham, *Annihilation of Positrons in Liquid Helium*, Phys. Rev. **106**, (1957) 16.

J. Wackerle and R. Stump, *Annihilation of Positrons in Liquid Helium*, Phys. Rev. **106**, (1957) 18.

B. K. Sharma, *Positron annihilation in solid krypton and xenon*, Appl. Phys. A **5**, (1974) 265.

(4) The time about 2...3 ns measured for the fluorescence extinction or the light scintillations in scintillators;

(5) The time interval about 2...3 ns from lighting an atomic lattice until the appearance of the first photoelectrons;

(6) The older experiments for determining by direct timing or interference the so-called "relaxation" time, also in the range $10^{-8} ... 10^{-9}$ s;

(7) The lifetimes around 10^{-9} s estimated for many unstable elementary particles whose decays are accompanied by emissions of significant amounts of energy, not only the light or heavy electrons, but even K mesons and long by about 1 m Λ, Σ, Ξ or Ω hyperons made by strong interaction, which is completely contrary to quantum physics of elementary particles:

"The trouble arises with the decay of the new particles. Their lifetimes range from about 10^{-8} to 10^{-9} seconds, which is on the time scale of weak interaction. But the particles are made, as we have seen, by strong interactions, the time scale of which is some 10^{-23} seconds. According to one of our most fundamental tenets – that of reversibility – a particle made in a strong interaction should also decay that way. (...) The only trouble is that they live 100,000 billion times longer than they should! It was this enormous discrepancy between their expected and observed lifetimes that was chiefly responsible for the designation "strange" or "queer" particles."[5];

(8) The numerous types of nanosecond accelerators and counters devised in the last half century, and so on.

These near lifetimes for so many elementary processes of entirely different nature can be explained only by a common dominant stage in their fulfillment, much longer than all the other different stages specific to each of them, and this common stage of all these radiative processes can only be the elementary act of photon emission.

Therefore, if the photon emission time of the ring electrons at relative rest $v \ll c$ is always $\tau_0 = 2.451 \cdot 10^{-9}$ s, and the beam photons radiated by them has the lower limit frequency $\nu_{min} = 4.08 \cdot 10^8$ Hz when they are made of the minimum number of preons $n_{min} = 2$, but the upper limit frequency $\nu_{max} = \nu_a = 1.234 \cdot 10^{20}$ Hz when they are made of all n_0 preons existing within a ring electron at rest, where $\nu_a = m_0 c^2 / 2h$ is the frequency of the two γ-photons resulted by electron-positron annihilation at rest, from the ratios $(n_0 - 1)/(n_{min} - 1) = \nu_{max}/\nu_{min}$ it results

$$n_0 = 3.025 \cdot 10^{11},$$

and then the preon mass

$$m_p = m_0/n_0 = 3.012 \cdot 10^{-42} \text{ kg}.$$

As for the size of the subquantum preons, it can be only estimated. If the smallest radii found for electrons from scattering experiments is below 10^{-19}, and the electrons accelerated in synchrotron to speeds $v \approx c$ reach masses equivalent even to 10^{16} preons, then the preons should have minimum dimensions about $10^{-19}/10^{16} \approx 10^{-35}$ m, hence somewhere near to the Planck's length $1.616 \cdot 10^{-35}$ m, anyway in a world which "may be a simple place, with just one kind of elementary particle and one important force", as several decades ago Georgi[79] concluded based on arguments of entirely different nature. Or, the world of preons even is one of some identical particles interacting only by magnetic force (besides universal gravity).

The one-dimensional beam photon, consisting in a variable number of very small subparticles evenly spread on their common direction of motion with the same velocity c, represents an unambiguous corpuscular model of this elementary particle, unlike the quantum pointlike photon, a corpuscle having mass m_γ, momentum $I_\gamma = m_\gamma c$ and energy $E_\gamma = m_\gamma c^2/2$, but omly *intrinsic* spin $L_\gamma = \hbar$, while the frequency v of its associated wave-train can have no connection with its energy E_γ, as long as the only part of the latter is to guide the motion of the corpuscle photon in substance, and the energy of a body is just a property of its mass.

Corpuscular beam photon instead wavicle photon

First corpuscular theory of the light was put forward by Gassendi at the middle of the 17th century, in 1690 Huyghens imposed a wave theory of the light, but in 1704 Newton reinforced its corpuscular understanding. Still this domination lasted only until the first decades of the 19th century, when the new-discovered phenomena of diffraction, interference and polarization brought back the supremacy of the wave theory, consolidated then in 1864 by Maxwell's electromagnetic theory conforming to which the whole electromagnetic radiation, including the light as a small segment of it, consists of

---

[79] H. Georgi, *A Unified Theory of Elementary Particles and Forces*, Sc. Amer. **244** (4), (1981) 48.

transverse waves whose electric $\bar{E}$ and magnetic $\bar{B}$ oscillating vectors are always perpendicular to each other and both to their propagating direction. And when Hertz discovered in 1886 the radio waves always transversely polarized, Maxwell's theory was considered to be definitely confirmed.

But yet, in 1905 Einstein subverted again this apparently immovable wave status of electromagnetic radiation, when he explained the recently discovered photoelectric effect by introducing the concept of light quantum, later called "photon". Nevertheless, as the concept of the pointlike photon could never explain the wave behaviors of this particle, as the own frequency, diffraction, interference, etc., but, on the other hand, many other experimentally noticed phenomena could never be explained without accepting its corpuscular nature, the photon became first "wavicle"[80] in the world of the elementary particles.

Finally, when in the modern age Maxwell's theory was clearly denied by discovering the longitudinal polarization of the light emitted by very fast particles accelerated in electromagnetic field, it was adopted the unifying concept of the pointlike photon guided by a wave train with a macroscopic finite length, but whose nature and physical medium of propagation has remained until now an impenetrable mystery.

But now the beam photons of macroscopic length, made of identical subquantum particles evenly spread on a straight line of determined length and coincident with their common direction of motion, have automatically all *undulatory* properties experimentally noticed to all linear fascicles of monochromatic electrons, protons, atoms, etc., and even much emphasized just because of their much higher regular structure made of subcomponents placed strictly on the same line at exactly equal distances between them, which amplifies very strongly their *undulatory* behavior noticed at macroscopic level, in comparison with those noticed in the same experimental conditions to the much less structurally ordered macroscopic beams of elementary particles or atoms.

As for diffraction phenomena, the main responsible for changing all elementary particles into wavicles just because the electron and X-ray dif-

---

[80] A fanny name proposed "as a compromise" by A. S. Eddington in his book *The nature of the physical world*, Macmillan, New York and University Press, Cambridge, 1928 in order "to accept the mystery as a mystery", or, more concretely, to accept the simultaneous wave and corpuscular properties of the light.

fraction through thin foils lead to pictures similar enough, a common corpuscular mechanism of them is more than verisimilar, and such a sole mechanism can be based only on the Me^{n+}-ions certainly existent in any strafed region of a thin metal foil. Even if such a common mechanism for electron, positron, neutron, deuteron or X-rays diffraction is not possible now, because the quantum relativistic photon has neither electric charge nor dipole magnetic field, the beam photon made of magnetic preons can undoubtedly be included in a corpuscular mechanism of diffraction based on magnetic interactions, a mechanism common for all magnetic particles passed through a thin foil, from electrons to individual preons, no matter how small magnetic moment the latter have.

Also, the own spin of the preons can replace the integer spin artificially attached to the pointlike photon for explaining why and how electromagnetic radiation is differently polarized at the level of macroscopic observation, just because all the preons in a beam photon have always parallel spins through their emission mechanism itself.

On the other hand, another essential characteristic of the beam photon is the absence of any internal cohesion. Unlike the spinning ring electrons, where the distances between the neighboring preons are measured in 10^{-23} m units, small enough for a direct interaction between their dipole magnetic fields, the minimum distances between two consecutive preons in the linear beam photons emitted by slow ring electrons correspond to the wavelength $\lambda_{min} \approx 10^{-12}$ m of the γ-photons resulted from (e^-e^+) annihilations at rest, and consequently the preons in all beam photons are entities completely isolated from each other, which fly inertially in vacuum without being able to interact between them.

That is why the ring electrons are indestructible entities, which can never be divided in smaller parts, while the beam photons emitted continuously in Universe can easy be broken into smaller and smaller fragments when they encounter other free or bound elementary particles, finally even to individual preons roaming chaotically in cosmic spaces, and consequently we are entirely justified in seeing the known macrocosm as a more or less isotropic bath of preons. And even if these subquantum particles are individually undetectable at macroscopic level, their doubtless existence in unbound state can naturally justify two essential questions much debated in modern cosmology:

(1) The mystery of the so-called "dark matter", whose experimentally noticed properties agree very well with those of preons:

— only one kind of particle, with exceedingly small size $d \approx 10^{-35}$ m and mass $m_p \approx 3 \cdot 10^{-12}\, m_0$, where m_0 is the rest mass of electron, therefore exactly the lower limit of the mass estimated in the last decades for the so-called *invisible axions*[81];

— inert and gravitational mass, therefore gravitational interaction;

— very high transparency for other electromagnetic radiation, suggestively noticed at macroscopic level when two beams of light intersect with one another without changing perceptibly their intensity, section, direction and velocity of motion;

— non-uniform distribution in Universe, higher inside the galaxies and clusters or in their vicinity, smaller in the intergalactic spaces.

In addition, as the universal preons interact magnetically with each other at distances below 10^{-23} m, therefore in the range of the effective action radius of the universal weak interactions, this coincidence can be seen as an additional argument for the older supposition about a possible magnetic nature of the weak interactions, as long as the weak interaction of the dark matter has been noticed experimentally.

(2) The Hubble redshift, together to all its controversial details, since in a universal bath of chaotically roaming preons, the linear beam photons, made also of free preons included yet in an ordered structure, cannot avoid a slow, but continuous and statistically uniform rarefaction by preon-preon scattering during a long travel in cosmic spaces. Or, for a terrestrial observer such a process is equivalent to a redshift directly proportional to the covered distance, but involving neither a Doppler broadening of the light emitted by far off cosmic sources, nor diffuse images of the farthest galaxies or clusters of galaxies, therefore exactly the two effects whose non-existence in experimental observations has always been an ultimate argument against any non-relativistic interpretation of the Hubble redshift. Once again a relativistic fancy proves to have a much simpler classical explanation.

---

[81] Hypothetical subquantum particles having gravitational and electromagnetic interactions between them, and considered responsible for the weak interactions between all elementary particles and the strong interactions inside the hadrons. They also change into and from photons in strong magnetic fields. Well, all these properties of the axions are totally found at the preons (see also the next chapter).

Moreover, the total absence of the internal cohesion of the photons is clearly proved by the noticed partition of one and the same photon by two electrons in two atoms in touch, possible even when the latter are differently excited, the only condition being the equality between the sum of the two equal or different excitation energies and the energy of the single photon absorbed *in common* by the two atoms[82]. Or, as each of these two different excitations needs an excitation photon with exactly determined frequency, different from the photon frequency necessary for the other excitation process, such simultaneous excitations by a single photon shows a selective absorption of the incident photon, possible only for a composite photon like the beam photon, but absolutely impossible for the pointlike photon of quantum relativistic physics, guided or not by a wave-train.

The Raman effect, consisting in small selective impoverishments Δv in the frequency of the photons reflected by certain substances, equal to frequencies v_a absorbed in different electronic transitions within the atoms or molecules encountered by the incident photons, $\Delta v = v_a$, can also be understood only by a partial and selective absorption of the incident photons.

More, as for evident reasons a beam photon can be absorbed by an atomic electron only if its own frequency v_γ is equal to the frequency v_{orb} of the orbital motion of the electron, we have so a new explanation for the fact that the atoms absorb only some few frequencies in the white light and are entirely transparent for all the other frequencies, as all absorption spectra are obtained. And this restriction has evidently to be also valid when the excited atoms re-emit in all directions the previously absorbed frequencies, when similar emission spectra are recorded. As known, even today this selectivity in absorbing the photons by atoms is *solved* in the most rudimentary manner, simply by postulating the existence of some "allowed" atomic states having "allowed" atomic orbits, terms empty of any physical significance.

And indeed, there is a special atom where this primary conditioning $v_\gamma = v_{orb}$ is experimentally demonstrated, namely the *geonium* atom[32]. This "artificial" atom has a single electron with quantified $n = 0, 1, 2, \ldots$ energy levels doubled by splitting in the external magnetic field, and with different velocity and orbital radius in each atomic state. Well, for any orbital

---

[82] M. H. Mittleman, *Excitation of two atoms by a single photon*, Phys. Lett. A **26**, (1968) 612.

transition $\Delta n = +1$, the only possible in geonium atom both without or with spin inversion, the electron in this atom has to absorb a photon whose energy $h\nu_\gamma$ is exactly equal to the quantity $2M_B B$, $h\nu_\gamma = 2M_B B$, where M_B is the Bohr magneton and B is external magnetic induction. Or, as $M_B = eh/4\pi m_0$ and the orbital radius of cyclotron motion is $r = m_0 v/eB$, where m_0 is the mass of the slow electron and v is its linear velocity, it results $2M_B B = h\nu_{orb}$, where $\nu_{orb} = v/2\pi r$ is the frequency of the orbital motion of the electron. Therefore, this experiment called geonium atom proves as clear as possible that an electron in orbital motion can absorb exclusively photons whose frequency is rigorously equal to that of its orbital motion,

$$\nu_\gamma = 2M_B B/h = \nu_{orb}.$$

Obviously only the beam photon made of subquantum particles evenly spread on a straight line almost coincident with their directions of motion can justify this singular photon absorption of the electrons in orbital motion, an experimentally proved reality quite incomprehensible for any other pointlike or composite photon.

Photon emission of differently polarized ring electrons

When a ring electron in uniform motion with linear velocity v emit a beam photon, all the radiated preons preserve this velocity vector after their detachment from the radiating ring electron, so that all of them remain permanently in the plane of the subquantum circular circumference of the latter, but move away from it with the linear velocity $v_i = c\sqrt{1 - v^2/c^2}$ they had on the subquantum circumference of the radiating ring electron at the moments of their successive detachments from one and the same point. In consequence, in the reference frame at rest the velocity vectors of all radiated preons in a beam photon is $\bar{v}_\gamma = \bar{v} + \bar{v}_i$, whose absolute value depends evidently on the constant angle between the velocity vectors $\bar{v}$ and $\bar{v}_i$ of the radiated preons at their detachment point from the radiating ring electron, therefore on the spin polarization of the latter.

There are two limit cases of spin polarization, 100 % longitudinal and 100 % transversal, while their mixture in variable proportions is called elliptical polarization of spin.

(1) The moving ring electron with longitudinal spin polarization has collinear spin $\bar{L}_e$ and linear velocity $\bar{v}$, and therefore the linear velocity $\bar{v}_i$

corresponding to the cyclotron motion of any preon on the subquantum circumference of the ring electron is every time perpendicular to the linear velocity $\bar{v}$ this preon also has, $\bar{v}_i \perp \bar{v}$, where $v_i = c\sqrt{1 - v^2/c^2}$.

The longitudinal polarization of the ring electrons is a very important case, because it includes not only all electrons accelerated in electromagnetic fields, from the electronic guns so much used in experimental research in the course of time, to the latest electromagnetic accelerators, but also the β-electrons and their precursors $\pi^{\pm} \to \mu^{\pm} \to e^{\pm}$, initially electrons strongly accelerated in electromagnetic field up to velocities very close to the speed of light c. First time the longitudinal spin polarization of the β-electrons was found by Lee and Yang[83], and then experimentally confirmed by many researchers, as well as for the electrons resulting from decaying muons.

As to the electrons accelerated in classical electromagnetic accelerators, at first known as "cathode rays" (Goldstein, 1873), their longitudinal spin polarization results directly when we take into account two simple facts: initially all the free electrons evenly spread on the surface of a negatively charged cathode have spins exclusively perpendicular to the surface and parallel between them as a result of the mutual interactions between their magnetic moments, and then these electrons are accelerated in directions also perpendicular to the cathode surface, as Goldstein and Crookes proved experimentally more than a century ago. In fine, in all accelerating electromagnetic fields, for example within the accelerating cavities in synchrotrons, the electrons have a longitudinal spin polarization, because the magnetic flux lines of these accelerating fields are always parallel to the motion directions of the accelerating photons and accelerated particles. Or, if in all electromagnetic fields the electrons are always accelerated in directions collinear with their spins, their longitudinal polarization results implicitly.

In consequence, when such a longitudinal ring electron with linear velocity $\bar{v}$ radiates in flight a linear beam photon, all the preons successively detached without mechanical recoil from one and the same point on its subquantum circumference have motion directions assessed by the vector sum $\bar{v}_\gamma = \bar{v} + \bar{v}_i$, where $v_i = c\sqrt{1 - v^2/c^2}$, and consequently they have always an invariant velocity c whatever velocity v the radiating ring electron has,

---

[83] T. D. Lee and C. N. Yang, *Question of Parity Conservation in Weak Interactions*, Phys. Rev. E **104**, (1956) 254.

$$v_\gamma = \sqrt{(v^2 + v_i{}^2)} = \sqrt{v^2 + c^2(1 - v^2/c^2)} = c,$$

and also the same emission angle β as against the velocity vector $\bar{v}$ of the radiating ring electron,

$$\beta = \arccos(v/c).$$

And indeed, these two equations $\bar{v}_\gamma = c$ and $\cos\beta = v/c$ deduced for the photon emissions of the ring elementary particles with longitudinal spin polarization are clearly validated by very numerous experimental data well known for a long time (when experimental data in the works cited further are given as polar diagrams, their feasible verification has to take into account the emissions angles corresponding to the maximum intensity of the recorded emissions, because there is always a large angular divergence of any photon radiation owing to the numerous disturbing factors).

First systematic experimental data proving simultaneously the two above equations resulted from the older study on the bremsstrahlung radiated either by the electrons electromagnetically accelerated in electronic guns, or by the β–electrons ejected from instable nuclei, when these rapid electrons are braked in dense substance, a radiation whose speed c was found to not depend on the velocity v of the radiating electrons, but whose noticed directions of emission were perpendicular to their motion direction ($\beta \approx 90°$) only when $v \ll c$, and then progressively directed to forward as their velocities v increased more and more, exactly as the equation of angular divergence $\beta = \arccos(v/c)$ predicts:

"For electrons with low velocities ($v/c = 1/20$) the maximum intensity is achieved in the vicinity of $\theta = \pi/2$. For high velocities ($v/c = 1/3$, $1/4$) the maximal emission is along the motion direction of the electrons."[84]

Still until now the most accurate experimental confirmation of emission angle $\beta = \arccos(v/c)$ remains synchrotron radiation emitted inside electromagnetic cavities by ultrafast electrons $v = 0.9\ ...\ 0.99999999\, c$ [85], where already long ago it was empirically established an identical angular divergence $\sin\beta = \sqrt{1 + v^2/c^2}$, or $\beta \approx \sqrt{1 + v^2/c^2}$ for very small angles β.

Also, an unexpectedly accurate verification of the the same angular divergence $\beta = \arccos(v/c)$ can be observed in the polar diagrams where

---

[84] E. V. Spolski, *Atomic Physics*, Vol. I, GITTL, Moskow, 1951. (in Russian)

[85] E. M. Rowe and J. H. Weaver, *The Uses of Synchrotron Radiation*, Sc. Amer. **236**, (1977) 32.

it is given the measured intensity of photon radiation emitted by certain extragalactic objects around their directions of motion with very high velocities $v = 0.5 \ldots 0.998\,c$ [86]:

v/c	0.50	0.90	0.95	0.97	0.98	0.99	0.998
$\beta°$ - experimental	60	25.3	18.1	14.0	11.7	8.0	3.7
$\beta° = \arccos(v/c)$	60	25.3	18.2	14.1	11.5	8.1	3.6

Very significant, the two equations $v_\gamma = c$ and $\cos\beta = v/c$ are accurately confirmed in all particle-antiparticle annihilations, either in flight with high velocities, or at rest in magnetic field.

Thus, when a pair electron-positron annihilate at rest in an uniform magnetic field, where the two electrons have spins evidently parallel to the flux lines, it results two γ-photons moving always with the same velocity c on directions perpendicular to the field lines[87], therefore under the emission angles $\beta = 90°$, while the positronium atoms (e^-e^+) strongly accelerated in electromagnetic field up to the velocity $v = 0.5\,c$ annihilate in flight into two γ-photons having the same velocity c and an angle $2\beta = 120°$ between them[88].

In an identical manner of disintegration, the neutral pions π^0 with velocity $v = 0.7\,c$ decay in flight into two γ-photons under emission angles $\beta = \arccos(0.7) = 45°$ [31], while the neutral pions with velocities in a large domain $v = 0.5 \ldots 0.99977\,c$ decay in flight into γ-photons whose emission angles decreased as the velocity v of initial pions was higher, even up to $\beta \approx 0°$ for $v \approx c$, but having all the same invariant velocity c [89]. And when neutral pions π^0 with velocities $v = \sqrt{3}c/2$ and $v = \sqrt{8}c/2$ disintegrate in flight[90], the noticed experimentally emission angles between the motion di-

---

[86] L. I. Matveenko, *Apparent superluminal separation velocities of the components of extragalactic objects*, Usp.Fiz.Nauk **140**, (1983) 463. (in Russian)

[87] V. W. Hughes, S. Marder and C.S. Wu, *Static Magnetic Field Quenching of the Orthopositronium Decay: Angular Distribution Effect*, Phys. Rev. **98**, (1955) 1840.

[88] D. Sadeh, *Experimental Evidence for the Constancy of the Velocity of Gamma Rays, Using Annihilation in Flight*, Phys. Rev. Lett. **10**, (1963) 271.

[89] T. Alväger, F. J. M. Farley, J. Kjellman, L. Wallin, *Test of the second postulate of special relativity in the GeV region*, Phys. Lett. **12**, (1964) 260.

[90] A.G. Carlson, J.E. Hooper and D.T. King, *Nuclear transmutations produced by cosmic-ray particles of great energy.— Part V. The neutral mesons*, Phil. Mag. Series 7 **41**, (1950) 701.

rection of the two resulting photons were $\beta = \arccos\left(\sqrt{3}c/2\right) = 30°$ and, respectively, $\beta = \arccos\left(\sqrt{8}c/2\right) = 20°$.

Worth remembering, although these close similarities between the decay of the neutral pions and those of the electron-positron pairs, both at rest and in flight with high velocities, prove clearly that the neutral pion is also an instable particle-antiparticle pair, this experimentally evident reality was finally denied yet for the sake of the quantum theories, even if initially it was without hesitation enunciated[91].

As for the applicability in annihilation processes of the two equations $v_\gamma = c$ and $\beta = \arccos(v/c)$ deduced for the usual photon emissions of longitudinal ring particles, it must be taken into account that at least in principle the latter are also photon emissions, the only difference between them and the trivial photon emissions being the mass quantity radiated outside in the form of photons by the involved elementary particles, partial in trivial photon emissions and total in particle-antiparticle annihilations.

(2) The moving ring electrons with transversal spin polarization have their spin and linear velocity permanently perpendicular to each other, $\bar{L}_e \perp \bar{v}$, and therefore any of its constituent preons has permanently coplanar velocity vectors $\bar{v}$ and $\bar{v}_i$, but at a certain moment the angle θ between these two velocity vectors of a certain preon can have any value $0 \ldots \pi$, depending on the point on the subquantum circumference where the considered preon is placed at that moment. For this reason, both the scalar value v_γ of the linear velocity of a beam photon radiated by a transversal ring electron and its emission angle β (between $\bar{v}_\gamma$ and $\bar{v}$) depend on this angle θ, so that when a big number of transversal ring electrons with the same velocity velocity v emit monochromatic beam photons whose identical frequency is ν, an observer at rest notices a relative spectral width $\Delta\nu/\nu$ of this monochromatic radiation depending on the velocity v of the emitters,

$$\Delta\nu/\nu = (\nu_{max} - \nu_{min})/\nu = 2v/c ,$$

with a maximum frequency $\nu_{max} = \nu(1 + v/c)$ for $\theta = 0$, when the beam photons are emitted "forward" ($\beta = 0$) and their velocity is also maximum,

$$v_{\gamma_{max}} = v_i + v = c\sqrt{1 - v^2/c^2} + v ,$$

---

[91] E. Fermi and C. N. Yang, *Are Mesons Elementary Particles?*, Phys. Rev. **76**, (1949) 1739.

and a minimum frequency $v_{min} = v(1 - v/c)$ for $\theta = \pi$, when the beam photons are emitted "back" ($\beta = \pi$) and their velocity is also minimum,

$$v_{\gamma_{min}} = v_i - v = c\sqrt{1 - v^2/c^2} - v.$$

Perhaps the best example of photon radiation emitted by ring electrons with transversal spin polarization is that of the electrons in cyclotron motion with constant linear velocity $\bar{v}$ after leaving one of the accelerating electromagnetic cavities in synchrotrons, when their spins $\bar{L}_e$ align to the external uniform magnetic field, and consequently they become perpendicular to $\bar{v}$. That is why, while inside the electrmagnetic cavities the electrons radiate photons with invariant velocity c and only in directions placed on the surface of a cone with solid angle $2\beta = 2\arccos(v/c)$ and symmetry axis coincident any moment with the motion direction of the emitters, when their photon emission keeps on a very short time even after leaving the electromagnetic cavities, they emit photons in very different directions but only in the plane of their cyclotron motion.

(3). Still the most electrons in nature, either free or bound in atoms, have intermediary angles $0 < \alpha < \pi/2$ between their spin $\bar{L}_e$ and linear velocity $\bar{v}$, hence an intermediary spin polarization, somewhere between 100% longitudinal and 100% transversal.

In this case, a very simple calculation gives the following formula the maximum and minimum velocities of the emitted photons:

$$v_\gamma = c\sqrt{1 \pm 2v/c \cdot \sqrt{1 - v^2/c^2} \cdot \sin\alpha},$$

whose simplified form for velocities $v \ll c$ of the radiating electrons is

$$v_\gamma = c \pm v/c \cdot \sin\alpha,$$

so that when a big number of ring particles with intermediary spin polarization $0 < \alpha < \pi/2$ and velocity v emit a monochromatic photon radiation v, an observer at rest finds a relative spectral width $\Delta v/v$ of this radiation depending both on the velocity v of the emitters and on their intermediary spin polarization defined by the angle α,

$$\Delta v/v = (v_{max} - v_{min})/v = 2v/c \cdot \sin\alpha.$$

This last formula should be applied especially to atomic electrons with known linear velocities $v \ll c$, whose spin polarization in their orbital motions in all atomic states could be determined from the measured spectral widths of the atomic spectral lines emitted by them.

As a conclusion of this section, the beam photons have always velocities and emission angles strictly dependent on the velocity and the spin polarization of their annular emitters, excepting the invariant velocity c of the photons emitted by longitudinal ring particles, in the main electrons either accelerated in electromagnetic field or ejected from unstable nucleons. In addition, conforming to the beam photon model the "natural" broadening of all spectral lines emitted by free or bound ring particles is an effect caused by the different velocities of the photons emitted by them in different directions and the different spin polarizations of the radiating ring particles, and not a consequence of the quantum uncertainty. For instance, the hydrogen maser microwaves $\lambda = 0.212$ m with no natural width can be understood now as being radiated by electrons with 100 % longitudinal spin polarization, and not by that ridiculous equalization of the lifetime of the excited state responsible for their emission with the staying time of their emitters inside the maser quartz bulb, arbitrarily established to be of about $1\,s$.

Polarized beam photons

Polarization of light was known and experimentally studied for a long time, and excepting the Newton's epoch this phenomenon was understood and explained exclusively in undulatory terms. Still in the last century this perception shifted gradually to its corpuscular understanding. Already in the Maxwell's theory this property of electromagnetic radiation was connected with two imaginary vectors added to electromagnetic waves, one electric $\bar{E}$ and the other magnetic $\bar{M}$, but because the electric and magnetic fields can be generated only by elementary particles, it follows that in fact Maxwell unwittingly placed the origin of polarization phenomenon in an indefinite corpuscular zone. Also, ever since 1929 Landé drew attention to the advanced analogy between the polarization of electromagnetic radiation and that of the newly discovered spin of electron. And because in the after years this similitude between the spin polarization of the electrons and the polarization of light has become increasingly clear, finally the wave photon of quantum physics has been endowed with an intrinsic angular momentum $\bar{L}_\gamma = \hbar$, whose orientation as against its velocity establishes its polarization:

(1) linear, when $\bar{L}_\gamma$ has a transverse orientation, perpendicular to its propagation direction;

(2) circular, dextrorotatory or leftrotatory, when $\bar{L}_\gamma$ has a longitudinal orientation, parallel or antiparallel to its velocity vector;

(3) elliptical, when $\bar{L}_\gamma$ has an intermediary orientation as against the photon velocity, between 100 % transverse and 100 % longitudinal.

As seen, in quantum relativistic mechanics this phenomenon of polarization proper to electromagnetic radiation already has an entirely corpuscular meaning, the wave train associated to each pointlike photon determines only its direction of propagation and its wave behavior in substance, but has no connection with its polarization. A big step back as compared to Maxwell's theory, indeed.

On the other hand, however, this quantum relativistic theory of photon polarization has right from the start the absolutely unacceptable disadvantage to infringe assumedly the universal law of angular momentum conservation: for example, a free electron can emit or absorb successively however much photons, each of them with the same spin $\bar{L}_\gamma = \hbar$, but without changing its spin $\bar{L}_e$, and an atomic electron in fundamental state $n = 1$ can directly jump, say, on the $n = 4$ orbit by absorbing only one photon with spin $\bar{L}_\gamma = \hbar$, but it can come again on its fundamental orbit in three successive stages, $(n = 4) \to (n = 3) \to (n = 2) \to (n = 1)$, each of them accompanied by a photon emission; or, inversely, an atomic electron can jump from $n = 1$ orbit to $n = 4$ by absorbing consecutively three photons, each of them with spin $\bar{L}_\gamma = \hbar$, but can come again on its fundamental orbit in one stage by emitting only one photon with spin $\bar{L}_\gamma = \hbar$.

Obviously this simplistic understanding of photon polarization in quantum physics cannot be kept any more at the beam photon made of corpuscular preons simply because such a structure has no internal motion, and consequently it cannot have spin and spin magnetic moment classically defined, like any spinning ring particle. But yet, a polarization of any beam photon can be defined by taken into account the ever identical spin polarization of its constituent preons, as long as all the preons in a beam photon have always parallel spins $\bar{L}_p$, and implicitly the same angle between their spins and their directions of motion, therefore the same spin polarization. Accordingly, a linear beam photon is 100 % linearly polarized if all its constituent preons have 100 % transversal spin polarization as against their velocity vectors, or it is 100 % circularly polarized if all its constituent preons have a 100 % longitudinal spin polarization as against their propagating direction, and between these two limit cases we have beam photons elliptically polarized, neither entirely circular nor entirely linear. Of course, a validation of this version needs compulsorily a good consonance between its predictions and experimental reality.

As all the preons constituent of a ring electron have always a linear spin polarization in their subquantum space, where they have a uniformly circular motion in a plane perpendicular to their spins, and therefore any minute their spins $\bar{L}_p$ are perpendicular to their linear velocity $\bar{v}_i$ on the subquantum orbit, $\bar{L}_p \perp \bar{v}_i$, they can keep this linear spin polarization even after they leave the subquantum circumference of the radiating ring electron and form a linear beam photon in inertial motion, but only when (1) the radiating ring electron has 100 % transverse spin polarization, or (2) the radiating ring electron has a linear velocity v negligible as compared to that of light c, $v \ll c$, so that the angle between the axis of the radiated beam photon and its velocity vectors $\bar{v}_\gamma$ is too small to have discernible consequences on the spin polarization of the radiated preons regardless of the spin polarization of the radiant ring electron. Examples in this respect are the linearly polarized γ-photons resulted from electron-positron annihilations at rest, and, respectively, the linearly polarized photons emitted in synchrotron by the electrons when they keep radiating a very short time even after leaving the accelerating electromagnetic cavities.

Differently from the ring electrons with 100 % transverse spin polarization, which radiate always only beam photons linearly polarized for evident geometrical reasons, the ring electrons with 100 % longitudinal spin polarization radiate beam photons whose degree of circular polarization increases directly proportional to the ratio v/c of their emitters, just because the emission angles β of the latter decrease conforming to equation $\cos \beta = v/c$ as their velocity v increases, and so the angles $(\pi/2 - \beta)$ between the $\bar{L}_p$ spins of all their constituent preons and the velocity vector $\bar{v}_\gamma$ have to decrease proportionally. And indeed, this prediction is always experimentally confirmed, from antenna radiation 100 % linearly polarized in a plane perpendicular to the alternating electric current that generated it, to the bremsstrahlung radiated by β-electrons or electrons electromagnetically accelerated from velocities $v \ll c$ up to $v \approx c$, whose degree of circular polarization is always equal to velocity ratio v/c of their longitudinal emitters. As a matter of fact, just the almost 100 % circular polarization of Tscherenkow radiation emitted forward by electrons and protons with high velocities $v \approx c$ was first experimental flat denial of Maxwell's theory.

Probably synchrotron radiation offers the clearest experimental evidence for all previous predictions concerning the photon emission of the ring particles. In truth, if inside the accelerating cavity of the synchrotron the accelerated electrons having longitudinal spin polarization radiate rigorously under emission angles $\beta = \arccos(v/c)$ and the photons emitted by them exhibit a degree of circular polarization exactly equal to their velocity ratio v/c, if the same electrons keep radiating even after leaving this cavity, when they begin moving on a circular trajectory perpendicular to the external uniform magnetic field and consequently their spin polarization changes instantaneously from 100 % longitudinal inside the cavity into 100 % transverse owing to their spin alignment to the flux lines of the external magnetic field, they emit only on directions in their orbital plane and all the radiated photons are 100 % linearly polarized[92].

As seen, the spinless beam photon made up of spinning subquantum preons has a polarization entirely predictable if its emitting ring particle has a known velocity and spin polarization, which is out of the question for the pointlike photon emitted by the pointlike electron. More, this spin transfer from the whole photon to its constituent preons has the big advantage to not entail evident infringements of the universal law of angular momentum conservation in all elementary acts of photon emission, as it happens in the case of the quantum relativistic photon with intrinsic spin $\hbar$ emitted by a pointlike electron with intrinsic spin $\hbar/2$.

Another very important benefit of this version of photon polarization is its ability to calculate exactly both the velocity and the spin polarization of all elementary particles that emit spectral lines with measurable polarization and spectral width, for instance those emitted by atomic electrons and nuclear nucleons. Such data would be extremely useful especially for building

---

[92] Unfortunately, although these emissions of linearly polarized photons in all directions in the plane of the cyclotron motion in synchrotrons have been experimentally noticed for long, and although quantum relativistic physics cannot explain how the momentum conservation is observed in these photon emissions in very different directions, the velocity of these photons emitted even forward or back by monochromatic electrons have never been measured, or at least comparatively estimated one way or other. And this lack of interest is more than curious, since these photon emissions of some monochromatic electrons in very divergent directions could offer the best experimental proof for the relativistic invariance of the velocity of light.

up a new atomic and molecular physics after reconsidering the numerical value of the Planck's constant, because the spin polarizations of atomic electrons show exactly how their orbits are placed in the magnetic fields within the atoms and nuclei.

Spiral neutrinos radiated through multiphoton emission

The equality $T_a = T_{rot} = 8.103 \cdot 10^{-22}$ s between the period T_a of the two γ-photons resulted from electron-positron annihilation at rest and the rotation period $T_{rot} = 2\pi r_e/c$ of the ring electron at rest proves that the rotation period T_{rot} is for the spinning ring electrons the smallest possible period of photon emission, and this fact is certainly in connection with the previously sketched emission mechanism of the linear beam photons by the motionless ring electrons. But for the same reasons we should also consider for all the moving ring electrons a similar lowest period of photon emission, even if the rotation period of the moving ring electrons increases as their linear velocity v increases, $T_{rot} = 2\pi r_e/c \sqrt{1 - v^2/c^2}$. If so, an elementary act of photon emission with an emission period below $T_a = T_{rot} = 2\pi r_e/c$ is impossible for all ring electrons, or, in other words, no ring electron can emit a photon whose energy is higher than its rest energy E_0.

There are no experimental data able to confirm or infirm directly this conclusion proper to the classical model of spinning ring electron, however, fact is that (1) γ-photons with energy higher than the rest energy of the light electrons are never noticed in the photon emissions of the latter, no matter how high energy these radiating light electrons have, such γ-photons can be emitted only by mesons or nucleons, and (2) the electron-positron pairs in motion annihilate sometimes into 2, 3, 4 or more γ-photons with equal energy, and the cause of these multiphoton annihilations could be just their total energy higher than their energy at rest.

Also, if in the most cases the light electrons radiate photons with frequencies varying from those of the microwaves to those of the γ-photons, when the electrons are accelerated in electromagnetic field to very high velocities, for example up to $v = 0.999999\ c$, and then these high-energy electrons are violently braked in just some tenths of nanosecond to much smaller velocities, they have to get rid of a mass about thousand times higher than their rest mass E_0 in an exceedingly short time, which could be possible only by radiating simultaneously a big number of γ-photons, each of them with energy equal to this rest energy E_0 of the electron. Therefore, in such cases we should take into account a multiphoton emission of the high-energy electrons, that is, an emission of many, or even very many photons with energy E_0 in only one elementary act of photon emission. Likewise, as the decays $\pi^{\pm} \to \mu^{\pm} \to e^{\pm}$ of the unstable heavy electrons raise similar problems even more acute, and these decays involve always the appearance of neutrinos or antineutrinos, and not of γ-photons, we already have some arguments to consider the neutrinos and antineutrinos as being entities resulting from such processes of multiphoton emission.

On the other hand, in principle a multiphoton emission of the high-energy spinning ring electron with longitudinal spin polarization cannot differ as mechanism from the usual photon emission of any longitudinal ring particle, already convincingly confirmed by the good agreement between its predictions and experimental data. And the simplest variant would be to consider neutrino as a *bunch* of many, or even very many γ-photons radiated by one ring electron in a single elementary act of photon emission, each of these γ-photons simultaneously radiated having an energy equal to the rest energy of the radiating ring electron.

Summarizing, neutrinos could be elementary particles emitted exclusively by the light or heavy ring electrons with longitudinal spin polarization, when they have to eliminate in a single elementary act of photon emission a mass and energy much larger than the rest mass and energy of the final ring particle, which could be possible only if these high-energy electrons, muons or charged pions radiate simultaneously many or very many beam photons whose energy is limited to the rest energy of the radiating ring particles, of course through an emission mechanism similar to that in the trivial case of a single photon emission. Therefore, all of the beam photons emitted in the course of a multiphoton emission leave tangentially the subquantum orbit of the radiating longitudinal ring particle from different points on the subquantum circumference of the latter and at the starting moments properly delayed to each other, so that, on the whole, the beam photons radiated in an elementary act of multiphoton emission go innertially along radial directions symmetrically disposed with respect to the subquantum circumference of the radiating longitudinal ring electron, whose angles β as against the velocity vector $\bar{v}$ of the annular emitter are extremely small, $\beta = \arccos(v/c) \approx 0$, since $v \cong c$. Also, as all the photons are emitted by longitudinal ring electrons with very high velocities, they move on directions almost perpendicular to their axis whereon all their preons are equidistantly placed, so that an observer located on their direction of motion cannot detect their frequency and consequently cannot realize their existence itself, as it happens when the motion directions of the beam photons make small angles with their axes.

What results in such an elementary act of multiphoton emission can be seen in the enclosed sketch, where the annular emitter is figured by a little central circle whose velocity vector $\bar{v}$ is perpendicular to the page plane, out of which four linear beam photons issue from four points symmetrically placed on its subquantum orbit, at starting moments successively delayed from each other with the time interval $T_{rot}/4$. All the four beam photons are figured only by their first three preons, numbered in the order of leaving the subquantum circumference of the radiating ring particle at time intervals equal to rotation period T_{rot} of the latter. And similarly to any usual uniphoton emission of a longitudinal ring particle, all the preons in a radiated beam photon are evenly

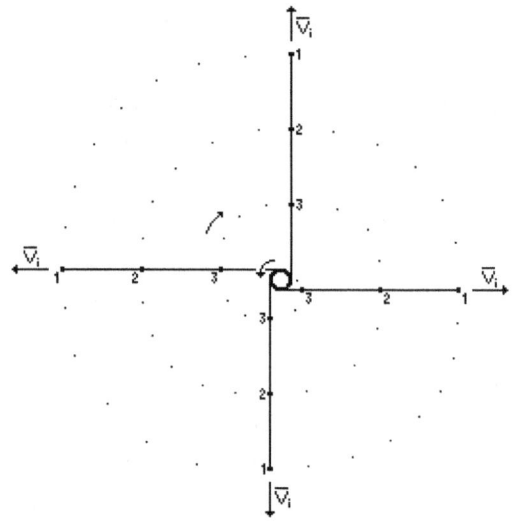

spread on a line tangent to the subquantum circumference of the moving emitter at the point of their successive detachments from the latter, a line along of which all of them have the linear velocity $\bar{v}_i$ they had on the subquantum circumference of the radiating ring particle exactly at the moments of their detachment from the annular emitter. In consequence, all the beam photons radiated in an elementary act of multiphoton emission remain lastingly coplanar with their annular emitter, because all their constituent preons keep even after their detachment from the radiating ring particle the velocity vector $\bar{v}$ perpendicular to the moving plane including permanently all of them, but in this plane all the radiated beam photons move constantly away from their annular emitter with linear velocity $v_i = c\sqrt{1 - v^2/c^2} \ll c$

(because always the linear velocity v of the latter is very high, $v \approx c$), on symmetrically divergent radial directions coincident with their axes.

Well, if all the radiated preons in the beam photons emitted quasi-simultaneously in an elementary act of multiphoton emission are joined by a dotted line in the global order of their detachment from the radiating ring particle, it results a spiral, therefore exactly that shape assigned long ago to neutrinos, in unison, but yet independently, by Lee and Yang, Landau or Salam (see a prompt review[93] of their almost simultaneous works[94]). And even if the spiral structure of neutrinos and antineutrinos was later forgotten by quantum relativistic theory of elementary particles (because how could a pointlike lepton have a spiral shape?), its unexpected update by the classical models of spinning ring electron and linear beam photon is extremely significant.

However, a final acknowledgment of this mechanism of multiphoton emission proper to the high-energy ring particles needs evidently an ampler verification of the consonance between the properties of these spiral structures radiated by the longitudinal ring particles of very high energy and the main properties of neutrinos, as they were experimentally noticed.

The most salient characteristics of this preonic neutrinos radiated by multiphoton emissions can easily be noticed in the enclosed sketch and then compared with all experimental data known in the field:

(1) As any neutrinic spiral is made of beam photons whose energy is equal to the rest energy of the radiating ring particle, which spring from different points on the the subquantum circumference of the emitting ring particle and then go inertially in directions tangent to their annular emitter with linear speed $v_i = c\sqrt{1 - v^2/c^2} \ll c$, and each preon in a spiral neutrino has a velocity vector given by the vector sum of its velocity vector $\bar{v}_i$ wherewith it departs from the radiating ring particle and the linear velocity vector $\bar{v}$ of the latter, $\bar{v}_\gamma = \bar{v}_i + \bar{v}$, the total scalar velocity of each radiated preon, and

---

[93] B. S. Rodberg and V. F. Weisskopf, *Fall of Parity. Recent Discoveries Related to Symmetry of Laws of Nature*, Science **125**, (1957) 627.

[94] T.D. Lee and C.N.Yang, *Parity Nonconservation and a Two-Component Theory of the Neutrino*, Phys. Rev. **105**, (1957) 1671.

L.D. Landau, *On the conservation laws for weak interactions*, Nucl. Phys. **3**, (1957) 127.

A. Salam, *On parity conservation and neutrino mass*, Nuovo Cim. **5**, (1957) 299.

consequently of all the beam photons constituent of a spiral neutrino is always $v_\gamma = \sqrt{(v^2 + v_l^2)} = \sqrt{v^2 + c^2(1 - v^2/c^2)} = c$;

(2) As only the ring particles electromagnetically accelerated to very high velocities $v \approx c$ can have multiphoton emissions, their emission angles as against their motion directions are very small, $\beta = \arccos(v/c) \approx 0$, therefore the spiral neutrinos and antineutrinos are emitted always *forward*;

(3) All the preons in the beam photons constituent of preonic neutrinos or antineutrinos have always almost 100 % circular polarization (or, more exactly, their degree of circular polarization is equal to the velocity ratio v/c of their annular emitters, but in principle this ratio is $v/c \approx 1$). And because the spin polarization of the preons can be either dextrorotatory or leftrotatory, preonic neutrinos have four distinct varieties in total: two with left-screw spirality and dextrorotatory or leftrotatory circular spin polarization, two with right-screw spirality and dextrorotatory or leftrotatory circular spin polarization.

As seen, if in quantum physics the number of distinct neutrinos has been a controversial problem for a long time, the classical model of preonic spiral neutrino gives the right answer right from the start;

(4) From the above sketched mecanism of multiphoton emissions one is seeing that the spiral preonic neutrinos radiated by different ring particles are to have different *patterns*, which differ between them because of the different radii of their emitters. For instance, a spiral neutrino radiated by a longitudinal ring muon with radius $r_\mu = r_e/207 = 1.868 \cdot 10^{-15}$ m and rotation period $T_{rot} = 2\pi r_\mu/c$ is right from the start 207 times more compact than a spiral neutrino made of the same number of beam photons and preons, but radiated by a longitudinal ring electron whose radius $r_e = 3.866 \cdot 10^{-13}$ m and rotation period $T_{rot} = 2\pi r_e/c$ are 207 times larger. In consequence, the spiral neutrinos radiated by the light or heavy ring electrons should inevitably have the distinctive mark of their origin, so that they are either compact muonic neutrinos v_μ, or more extensive electronic neutrinos v_e. Or, this different initial compactness of the spiral neutrinos radiated by different ring particles has evidently to lead to differences regarding their interactions with other near elementary particles, especially with nucleons, whose dimensions are much closer to that of the muons. And indeed, these two types of neutrinos were experimentally confirmed in 1961, just by their different way of reacting with nucleons;

(5) Any preonic neutrino experiences right from the beginning of its emission a continuous dilatation, because all its constituent beam photons move continuously away from the central radiating ring electron in their motion directions always tangent to the subquantum circumference of the annular emitter, with the same linear velocity $v_i = c\sqrt{1 - v^2/c^2} \ll c$ they have on the subquantum circumference of the radiating longitudinal ring electron before their successive detachments from the latter. The higher the linear velocity v of the radiating longitudinal ring electron, the smaller the linear velocity v_i of the linear beam photons in directions coinciding with their axes, and implicitly the slower the spatial dilation of the spiral neutrino in the plane of the subquantum circumference of its annular emitter.

And indeed, this more or less rapid spatial extension of spiral neutrinos can explain as simple as possible two very important questions in neutrino physics, namely the experimentally noticed differences between the "near" and "far" interactions of the just radiated neutrinos with other elementary particles, and why many have looked in vain for neutrinos certainly emitted by hot celestial bodies, for instance the Sun, which becomes easy to understand if we have in view the relatively rapid decay by an endless dilation of any spiral neutrino into a large number of individual γ-beam photons moving in different directions and anyway experimentally undetectable through their own frequency owing to the large angles between their axes and their motion directions. Moreover, it is well known that the frequency of all γ-photons emitted by longitudinal high-energy particles cannot be measured experimentally, and this reality cannot be due to an exceedingly high penetrability in substance of these photons, a supposition supported by no logical argument. Only the one-dimensional structure of the beam photons can explain it, namely by the fact that the preons of the beam photons whose motion directions make larger angles with their axes are received rather simultaneously by the macroscopic observers placed on their direction of motion, and so these observers cannot distinguish a reception frequency, the usual way by which they can identify the photons whose constituent preons are successively received because they are evenly spread on a straight line almost coincident with their motion directions.

In addition, this spatial dilation of the preonic neutrinos can explain why a surprising change of muonic neutrinos into electronic neutrinos $v_\mu \to v_e$ was experimentally noticed, but never an inverse change. As evident, such an one-way conversion $v_\mu \to v_e$ is definitively inexplicable for the pointlike neutrino of quantum relativistic physics, but it is a normal consequence of the spatial extension of any preonic spiral neutrino, because of which after its emission the initially much compact muonic spiral neutrino reaches at a certain moment the larger size of the electronic spiral neutrinos, and consequently it can react as the latter with nucleons.

As normal, the end of the spiral neutrinos radiated by high-energy longitudinal ring particles is that of the beam photons which they are made of, whose ulterior repeated fragmentation in substance supply finally the undetectable dark matter;

(5) Conforming to the enclosed sketch of multiphoton emission, the longitudinal light or heavy ring electrons positively charged, which have right-screw helicity (or rotation in the direction of clockwise), have to radiate always spiral neutrinos with left-screw spirality, and vice-versa. For instance, in the enclosed figure the central radiating ring electron spins in the opposite direction of clockwise, therefore it is a negative electron, while the radiated preonic spiral develops in the rotation clockwise, therefore it is an antineutrino;

(6) Anyone having a certain knowledge in neutrino physics can find a good agreement between all the above mentioned properties of the spiral neutrinos and antineutrinos emitted by longitudinal ring particles through their specific multiphoton emission and those experimentally noticed to these elementary particles, but also an essential difference between them and those proper to the pointlike neutrinos and antineutrinos in quantum relativistic physics, namely their spins.

In truth, since the spiral neutrinos have not (as a whole) an own rotation defined by an angular momentum and able to generate a dipole magnetic field defined by a magnetic moment, their own spin is evidently null, and therefore, exactly as in the case of the spinless beam photons from which they are made, their spin polarization is in fact the unique spin polarization of all their constituent preons. Or, from their emission mechanism itself it is easy to conclude the almost 100 % circular polarization of all spiral neutrinos or antineutrinos, just because they are always made of preons with almost 100 % circular spin polarization.

Differently, in quantum relativistic physics the pointlike neutrinos and antineutrinos have been endowed with an intrinsic semi integer spin $\hbar/2$, like the electrons and muons, although neutrinos and antineutrinos are elementary particles radiated by the latter like the photons, particles also without rest mass, but endowed by an intrinsic integer spin $\hbar$. Moreover, the history of the intrinsic spin of neutrinos and antineutrinos is very significant.

If the pointlike photons of quantum relativistic physics have been endowed with an integer spin $\hbar$ much after the experimental acknowledgement of the photons as distinct elementary particles, the semi-integer spin $\hbar/2$ was assigned to neutrinos by Pauli many years even before the experimental demonstration of their real existence. And not at all accidentally this hurried initiative came from the one who had some years before that brilliant idea of the dipole magnetic fields generated by the orbital motions of the atomic electrons, by which he removed from a hard impasse the Bohr's theory of the free hydrogen atom, an entity whose existence was assumed with no sound argument.

Why in 1930 Pauli was in a hurry to predict the existence within nuclei of a still unknown neutral particle with semi integer spin is not hard to understand: few years earlier experimental research found a nuclear spin variation $\Delta s = \pm n\hbar$ in all β-decays, where $n = 1, 2 \ldots$ is an integer number and $\hbar = 1.056 \cdot 10^{-34}$ J·s , a spin variation which could not be explained by the intrinsic spin $\hbar/2 = 0.528 \cdot 10^{-34}$ J·s assigned to the electrons by quantum physics, so that another entity with spin $\hbar/2$ was supposed to exist in nuclei, or at least to result together one β-electron in any nuclear decay of this kind, and finally this supplementary entity was identified to be the particle which also takes over some of the energy lost by initial nucleus during its β-decay, particle included in the table of elementary particles as a new constituent called neutrino.

Well, owing to this postulated semi integer spin $\hbar/2$ of neutrinos, as well as owing to the integer spin $\hbar$ assigned to charged pions by meson theory of nuclear interactions, now we have the following complete equations for the decaying processes $\pi^\pm \to \mu^\pm \to e^\pm$ of the charged mesons:

$$\pi^- \to \mu^- + \bar{\nu}_\mu \qquad \pi^+ \to \mu^+ + \nu_\mu$$
$$\mu^- \to e^- + \bar{\nu}_e + \nu_\mu \qquad \mu^+ \to e^+ + \nu_e + \bar{\nu}_\mu.$$

Notice, in quantum physics when the charged pions decay only the resulted muons emit neutrinos or antineutrinos, but when the muons decay both initial and final particles emit their own neutrinos or antineutrinos!

Differently, if the neutrinos and antineutrinos have null spin, the two equations of disintegration become coherent with one another:

$$\pi^- \to \mu^- + \bar{\nu}_\mu \qquad \pi^+ \to \mu^+ + \nu_\mu$$
$$\mu^- \to e^- + \bar{\nu}_e \qquad \mu^+ \to e^+ + \nu_e.$$

More, these new disintegration equations lead to a very interesting conclusion, because now each of them can be seen as a process in two stages: first time the unstable meson change into another particle by dilation, $\pi^\pm \to \mu^\pm$ and $\mu^\pm \to e^\pm$, and then the resulted particle radiates its mass in excess under the form of its characteristic neutrino, muonic or, respectively, electronic, $\mu^\pm \to \mu^\pm + \nu_\mu$ and $e^\pm \to e^\pm + \nu_e$. And because the two stages have certainly very different lengths of time, since initial dilations of the unstable ring mesons can last an extremely short time, the global lifetime of these charged mesons should be practically equal to the time necessary for the final neutrino emission in each disintegration.

On other hand, as any neutrino emission of a longitudinal high-energy ring particle is in fact a multiphoton emission, whose duration is equal to that of an usual photon emission, which in turn is quantified by the rotation period $T_{rot} = 2\pi r/v_i$ of the radiating ring particle, where $v_i = c\sqrt{1 - v^2/c^2}$ is the peripheral linear velocity of the rotating annular emitter, it results the general formula for the approximate lifetime τ of all unstable particles whose decay ia accompanied by a photon or multiphoton radiation,

$$\tau = \tau_0/\sqrt{1 - v^2/c^2},$$

where τ_0 is the lifetime of the same unstable particles at relative rest, or, more accurately, when their velocities are much smaller than the velocity of light, $v \ll c$.

The formula above is identical with that of the relativistic dilation of time, whose first claimed experimental confirmations[95] were referring exactly to the lifetime of the very rapid muons detected on the earth at very high

---

[95] B. Rossi, K. Greisen, J. C. Stearns, D. K. Froman and P. G. Koontz, *Variation of the Rate of Decay of Mesotrons with Momentum*, Phys. Rev. **59** (1941) 223 and

B. Rossi and N. Nereson, *Further Measurements of the Mesotron Lifetime*, Phys. Rev. **61** (1942) 675.

D. H. Frisch and J. H. Smith, *Measurement of the Relativistic Time Dilation Using μ-Mesons*, Am. J. Phys. **31** (1963) 342.

altitudes, and that of the slow muons detected at the sea level after their sharp braking in terrestrial atmosphere by repeated collisions with the gas molecules in the air.

From Newton's particles of light to preons

The beam photon made of preons and the Newton's vision on the light have some amazing similarities, which sometimes go even to details.

Thus, in his famous *Opticks*[55] Newton concluded that the light is made of very little but very fast particles, or "corpuscles", which have mass, momentum and kinetic energy, and therefore they must obey the same laws of physics as all the other masses noticeable at macroscopic level, for instance the planets. Among other things, Newton explained that the light appears to travel only in straight line just because of its very high velocity, exactly as the trajectory of the very fast balls seems apparently to be straight, a statement much later experimentally confirmed by the curvature of the luminous rays in the solar gravitational field.

But Newton's belief in corpuscular nature of the light was not at all a simple intuition, his conviction was based on observations whose significance has slipped the mind of most others, either before or even after him. For instance, one of his arguments in this respect was the banal fact that in a vacuum two luminous rays pass through each other without changing their motion direction, intensity, color or polarization. Or, it is evident that two waves, either luminous or of other nature, cannot intersect in space with no consequence, just because all the waves have by definition a continuous spatial extension and consequently their spatial superposition leads inevitably to a new spatial entity different in all respects from initial waves.

On the other hand, it is similarly evident that two corpuscular luminous rays can intersect with no perceptible consequence only if they are made of particles so tiny and so much distanced to each other, that only an absolutely negligible number of them get really to collide and scatter reciprocally. In a time when the atoms and their constituents were not known yet, Newton understood that the light must be an extremely rarefied matter in comparison with that one out of which all the bodies around him are made. And now the beam photon model is in full agreement with his reasoning, as long as the the distances between two successive preons in all beam photons are at least of about 10^{-12} m, therefore at least over 10^{20} times larger than the estimated size of the preons, whose dimensions are certainly below 10^{-35} m!

More, Newton provided not only the exceedingly small dimensions of his particles of light, but also their geometrical shape: indeed, in order to explain why some translucent substances can rotate the polarization plane of the light, Newton stated that the light particles cannot be spherical, they could have only a plain form, most probably as a plate. Correspondingly, for reasons concerning the high internal cohesion of the spinning ring electron, its constituent preons should have most probably an annular form similar to that of the ring electrons themselves, therefore a geometrical form close to a plate. And because in a beam photon all its constituent preons have always the same spatial orientation of their angular and magnetic moments with respect to their parallel directions of motion, and this unique orientation defines what polarization that photon has, this connection between geometrical shape of the light particles and all phenomena of light polarization, including the noticed rotation of the polarization plane of the light passing through certain translucent bodies, is also in agreement with the beam photon made of subparticles having most likely the flat shape of some rotating rings with parallel symmetry axes.

But maybe the most impressive of the Newton's thoughts about the light refers to its relation with macroscopic bodies, which can both emit and absorb it:

"Are not gross Bodies and Light convertible into one another, and may not Bodies receive much of their Activity from the Particles of Light which enter their Composition? For all fix'd Bodies being heated emit Light so long as they continue sufficiently hot, and Light mutually stops in Bodies as often as its Rays strike upon their Parts, as we shew'd above ...

The changing of Bodies into Light, and Light into Bodies, is very conformable to the Course of Nature, which seems delighted with Transmutations."

Is not this a foreknowledge of the full unity of matter, or of the Heisenberg's *universal matter* out of which all elementary particles are made, including the photons, long before discovering these elementary particles and their endless chain of mutual transformations?

FOUR FUNDAMENTAL INTERACTIONS

Universal weak interaction

If all elementary particles are made of identical preons with mass $m_p = 3.012 \cdot 10^{-42}$ kg and own magnetic moment, the experimentally noticed universality of the weak interaction should be connected to these universal subquantum particles, and therefore the weak interactions could be identified with their magnetic interactions, directly noticeable only when the distances between them are small enough. In other words, two elementary particles have a weak interaction when their nearest preons, at least one in each of the two particles, come to a distance smaller than the effective action radius of their magnetic interaction. Moreover, the reality of a weak interaction existing exactly at the contact point between two elementary particles has been acknowledged for a long time[96].

There are many other arguments which support this logical supposition, but maybe the most eloquent is the experimentally noticed dependence of the strength of the weak interaction between two elementary particles in contact on the mutual orientation of their spins, a dependence specific to mutual interactions between magnetic dipoles. Or, as the spin $\bar{L}_e$ of any ring particle is always parallel (or antiparallel) to all spins $\bar{L}_p$ of its constituent preons, whether this elementary particle is at rest or in motion, free or bound in more complex structures, a certain mutual orientation of the two spins $\bar{L}_e$ of two ring particles in contact means automatically the same mutual orientation of the two spins $\bar{L}_p$ of their two nearest preons.

Similarly, the spin polarization of the beam photons and spiral neutrinos made of beam photons is in fact the identical spin polarization of all their constituent preons.

The known inner structure of the light and heavy ring electrons afford even clear quantitative arguments supporting this new understanding of the universal weak interactions between all elementary particles.

---

[96] D. B. Cline, A. K. Mann and C. Rubbia, *The Detection of Neutral Weak Currents*, Sc. Amer. **231**, (1974) 108.

Thus, as the light ring electrons at rest have $n_0^e \approx 10^{11}$ preons, the distances between two preons neighboring on their subquantum cyclotron orbits is $d_e = 2\pi r_e/n_0^e \approx 10^{-23}$ m, which is indeed a typical action radius of the weak interactions. And within the ring muons or charged pions at rest the distances between two neighboring preons on their subquatum orbits are still much smaller, $d_\mu \approx 10^{-28}$ m and $d_\pi \approx 10^{-29}$ m. Of course, in the light and heavy electrons in motion with very high velocities the distances between their constituent preons is proportionally smaller as their mass is higher than their rest mass.

Also, as the preons are subquantum magnetic dipoles with axial symmetry, at these so small distances their magnetic interaction force has to be assessed by a law in $1/d$ (very different from that in $1/d^4$ valid at much larger distances), and in truth this law in $1/d$ has been found experimentally for the universal weak interactions.

In fine, the non-dimensional constants g_w of the weak interaction experimentally found for light and heavy electrons can be calculated from the known radii of the ring leptons, $r_e = 3.866 \cdot 10^{-13}$ m, $r_\mu = 1.868 \cdot 10^{-15}$ m and $r_\pi = 1.366 \cdot 10^{-15}$ m, starting from the Fermi beta decay constant G_F defined by the basic formula $G_F/(\hbar c)^3 = 1.166 \cdot 10^{-5}/\text{GeV}^2$, wherefrom it results $G_F = 1.437 \cdot 10^{-62}$ J·m³. From this value we get the weak interaction constants of electrons, muons and pions,

$$g_w^e = G_F/r_e^2 = 0.955 \cdot 10^{-37} \text{ J·m}$$
$$g_w^\mu = G_F/r_\mu^2 = 4.117 \cdot 10^{-33} \text{ J·m}$$
$$g_w^\pi = G_F/r_\pi^2 = 7.700 \cdot 10^{-33} \text{ J·m},$$

which on the QED scale $\hbar c = 1$ become the known non-dimensional constants for the weak interactions of these light and heavy electrons,

$$g_w^e = 3.019 \cdot 10^{-12}$$
$$g_w^\mu = 1.302 \cdot 10^{-7}$$
$$g_w^\pi = 2.435 \cdot 10^{-7}.$$

The same non-dimensional constants of the light and heavy electrons (valid on the scale where $\hbar c = 1$) can be obtained from the other formula proposed for the Fermi's constant, $G_F = 10^{-5}/m_p^2$, where the mass m_p of the proton is given in relative values $m_p/m_e = 1836.1$, $m_p/m_\mu = 8.870$ and $m_p/m_\pi = 6.726$,

$$g_w^e = 2.966 \cdot 10^{-12}$$
$$g_w^\mu = 1.271 \cdot 10^{-7}$$
$$g_w^\pi = 2.211 \cdot 10^{-7}.$$

The small differences between the non-dimensional constants g_w^e, g_w^μ and g_w^π calculated from the two formulas usually used for the Fermi constant, $G_F = 1.437 \cdot 10^{-62}$ J·m³ or $G_F = 10^{-5}/m_p^2$, are not a problem, because these two formulas are not quite rigorously deduced, so that both sets of non-dimensional Fermi constants can be considered in good agreement with experimental data in the field.

Perhaps the most illustrative example of weak interaction between two ring particles is that of a pair ring electron-ring positron whose two annular components are coplanar, tangent and with parallel spins, therefore with the same sense of rotation. Because their two preons in touch at a given moment (one in the ring electron and the other in the ring positron) have antiparallel magnetic moments, both of them experience at that moment a null magnetic induction, and consequently they pass simultaneously from their uniformly circular motion on the two tangent subquantum orbits to an inertial motion in opposite spatial directions. And if these simultaneous detachments of the two nearest preons take place at constant time intervals, equal to the rotation period of the two ring electrons, they result in forming two linear beam photons with opposite directions of motion, the same frequency and 100 % linear polarization. Well, this electron-positron annihilation has been proved indeed to be a typical weak interaction, but now the ring electron model offers an intuitive mechanism of this phenomenon.

An additional notice is required: the null magnetic induction in the contact point of the two coplanar ring particles presumes antiparallel magnetic moments of the two coplanar ring preons existing at the considered moment in that contact point, and this proves that the ring electron and the ring positron should have different orientations of the spins $\bar{L}_p$ of their constituent preons as against their $\bar{L}_e$ spins, parallel orientation $\bar{L}_p \uparrow\downarrow \bar{L}_e$ in first of them, antiparallel $\bar{L}_p \uparrow\downarrow \bar{L}_e$ in the second, which confirms these different orientations as a structural factor able to differentiate by itself the electrons from positrons, and generally the particles from their antiparticles.

Moreover, these different orientations $\bar{L}_p \uparrow\uparrow \bar{L}_e$ and $\bar{L}_p \uparrow\downarrow \bar{L}_e$ inside the ring electrons and the ring positrons could explain the reverse process of annihilation, namely the pair creation $\gamma \to e^- + e^+$. Here we have initially just one γ-photon made of preons with parallel spins $\bar{L}_p$, but if in the *creation* process, which is possible only within the huge magnetic field around a very small magnetic particle, for example a nucleon, for one reason or another half of the preons begin a cyclotron motion in clockwise, and the other half in the opposite sense, it results so two ring electrons with antiparallel spins $\bar{L}_e$, but with parallel spins $\bar{L}_p$ of all their constituent preons, therefore a ring electron and a ring positron.

As seen, these mechanisms of the ring electrons annihilation and creation can explain why only the electron-positron pairs can annihilate or create, but never the electron-electron or positron-positron pairs.

Also, now it becomes clear why only pairs of annular particles can annihilate, whether (e^-e^+), or $(\mu^-\mu^+)$, or even $(e^-\mu^+)$, but never neutrinos or antineutrinos, even if the latter have also weak interactions with other particles, and also why (e^-e^+) and $(\mu^-\mu^+)$ annihilations have different constants of weak interaction.

Two short comments before ending this section:

(1) If the weak interactions between elementary particles in contact are in fact magnetic interactions between their nearest preons, then at the subquantum level these magnetic interactions between very close preons has to be extremely strong, because only so they can have effects noticeable at macroscopic level, which affect elementary particles whose mass is about 10^{12} times bigger, or even more, than the mass of the two very close preons interacting through their dipole magnetic fields;

(2) The high strength of this magnetic interaction between the very close preons is also proved by the known indestructibility of the electrons, which can be disintegrated only in particle-antiparticle annihilations, therefore only through the same extremely strong magnetic interaction between two very close preons.

Strong interaction between coaxial ring quarks

If the weak interaction of two coplanar and tangent ring electrons at rest is actually the trivial magnetic interaction between their two nearest preons at the contact point, the only preons near enough each from other for a significant interaction between their dipole magnetic fields, if all constituent preons of two coaxial and very close ring electrons have parallel magnetic moments, the weak interactions between all $n_0^e = 3.025 \cdot 10^{11}$ pairs of *face-to-face* preons cumulate in a total magnetic interaction of $3.025 \cdot 10^{11}$ times stronger than the weak interaction between two ring electrons at rest. In other words, the non-dimensional constant of the total magnetic interaction between all the preons with parallel magnetic moments in two coaxial and near enough ring electrons at rest and becomes $g_s = n_0^e g_w^e \approx 1$ on the QED scale, just by summing all the $3.025 \cdot 10^{11}$ weak interactions (each of them characterized by the non-dimensional constant $g_w^e = 3.019 \cdot 10^{-12}$) acting between all their $3.025 \cdot 10^{11}$ pairs of nearest (*face-to-face*) preons!

Or, as $g_s \approx 1$ is exactly the non-dimensional constant of the strong interactions on the scale where $\hbar c = 1$, the old question "When the weak interaction becomes strong?"[97] gets a very simple answer: when two ring particles at distances smaller than the effective action radius of the weak interaction pass from a tangential and coplanar coupling to one coaxial.

This meaning of the strong interaction is clearly supported by the fact that the hadrons are created in laboratory at the collision point of two collinear and opposite beams of high-energy electrons and, respectively, positrons. Indeed, in these conditions the longitudinal ring electrons and positrons are simply forced to get very close to each other and to couple coaxially by the attractive magnetic forces between all their nearest preons with parallel magnetic moments, resulting in longer or shorter structures made of coaxial ring electrons and positrons not only strongly condensed, but also strongly contracted in the huge magnetic field generated together by them in a very small space.

As these structures are made of strongly contracted and simultaneously strongly condensed coaxial ring electrons and positrons bound to their neighbors through strong interactions, we can identify them as being hadrons made of coaxial quarks and antiquarks,

$$e^- + e^- \rightarrow q + \bar{q} \text{ (hadrons)}.$$

In principle, these hadrons should have inner structures with axial symmetry, made of coaxial ring quarks (in fact heavy ring electrons very

---

[97] D. I. Blokhintsev, *When the weak interaction becomes strong?*, Usp. Fiz. Nauk **62** (1957) 381. (in Russian)

strongly contracted), either of type $[qqq\bar{q}\bar{q}\bar{q}]$, or maybe even $[q\bar{q}q\bar{q}q\bar{q}]$. All the ring quarks and antiquarks in the hadrons should differ between them by their mass, radius and magnetic moments, because all these properties depend on magnetic induction experienced by each quark or antiquark, according to the place it has in such coaxial structures of different lengths. In contrast, all the ring quarks and antiquarks have always to have the same immutable spin $L_{q^\pm} = L_{\pi^\pm} = L_{\mu^\pm} = L_{e^\pm} = 1.0545 \cdot 10^{-34}$ J·s.

Obviously the inner structure of such hadrons cannot be rigid, made of motionless quarks and antiquarks, they are to have a dynamical inner equilibrium resulted from synchronized motions forward-back of all their constituent ring quarks along their common symmetry axis.

More, this dynamical inner equilibrium of the hadrons can only be provisional, so that when the amplitude of these internal oscillating motions increases over a certain limit in a point or other of a chain of coaxial ring quarks and antiquarks, this chain can break at that point resulting two fragments which fly in opposite directions placed on the symmetry axes of the unstable hadrons. And if only one ring quark or antiquark detaches from the chain, its *instantaneous* dilation change it into a longitudinal ring meson, probably a charged pion, $q^\pm \to \pi^\pm$, which decays further conforming to the known succession $\pi^\pm \to \mu^\pm \to e^\pm$. The higher magnetic induction where a contracted ring particle is stable, the faster its decay into the next state, and because the first distinct stage $q^\pm \to \pi^\pm$ of complete dilation $q^\pm \to e^\pm$ has certainly to be much faster than the followings, the inability to detect the ring quarks at macroscopic level is quite explainable. Also, as these first particles expelled from the unstable hadrons are to have very high velocities $v \approx c$ and longitudinal spin polarization, and then their decays produces in flight along a straight line collinear with their spins, all the other particles resulted through their subsequent successive decays are to have the same longitudinal spin polarization, so that all neutrinos and antineutrinos emitted in each stage of these successive transformations by dilation have to be practically 100 % circularly polarized.

Therefore, using two opposite beams of longitudinal high-energy electrons and positrons for synthesizing hadrons is necessary firstly in order to stop them violently at the point of their collision, while the different nature of the two beams, one made of electrons and the other of positrons, is necessary for ensuring the parallel orientation of all magnetic moments of all preons in the two kind of ring particles coming from opposite direction, the obligatory condition for an attractive character of all strong interactions which appear in these conditions among the coaxial new formed ring quarks. In the other case, when the two opposite beams would be made of the same kind of particles, either electrons or positrons, they could not give birth to hadrons just because two high-energy electrons or positrons coming from opposite direction cannot bind one another through strong interaction owing to the antiparallel orientation of the magnetic moments of their constituent preons.

As for dimensional estimations, in a rough first approximation the coaxial ring quarks within the hadrons should have radii somewhere in the range 10^{-18} ... 10^{-20} m , and synchronized oscillation motions along their common axis of symmetry on distances of magnitude orders not too much different. In addition, these oscillating movements should imply a continuous variation of their radii and magnetic moments, depending at each moment on the distance between the neighboring ring quarks, which means also a continuous variation of the attractive magnetic forces between them. The smaller this distance at a given moment, the smaller their radii, magnetic moments and attraction forces, so that sometimes the attractive magnetic force between two neighboring quarks can even increase when they move away from each other, of course, at distances below certain limits (as it was already discussed, a similar case is the neutron made of two coaxial ring particles, a positive proton and a negative meson).

On the whole, the hadrons made of coaxial ring quarks bound to one another by strong interaction could be somewhat compared with the flexible springs made of a very elastic and thin wire, in continuous pulsation along their symmetry axis, but the magnetic field responsible for the weak interactions among their coaxial ring quarks is not similar to that of a solenoid, but rather to that of a permanent magnet similar to a pipe with very thin walls.

All the aspects speculated above can be compared with some general characteristics experimentally deduced for the very numerous hadronic resonances synthesized in accelerators on the way $e^- + e^- \to$ hadrons , as for example those mentioned in some few excerpts from a work[98] which tries to translate experimental data in the field into an intuitive language:

---

[98] M. S. Marinov, *Relativistic strings and dual models of the strong interactions*, Usp. Fiz. Nauk **121**, (1977) 377. (in Russian)

"Now the inner structure of the hadrons is much more evident and concrete. The hadrons appear as extended objects, probably composed of a small number of *fundamental particles* called quarks, which carry quantum numbers, as the charge and strangeness, and are bound by their gluon vectorial fields. This picture partially remind of the system of positive and negative particles (for examples electrons and positrons) bound by electric field. Still neither free gluons nor the *photons* (gluons) of strong interaction have been discovered experimentally. (...) The gluon field does not spread in space like electric field (slowly enough, conforming to the law of decreasing the charge with the distance), but it concentrates in the narrow tubes which form the sources of the field — the quarks.

(...) The flat nature of dual charts indicates that the field carrying the interaction between quarks, for whatever reason, is concentrated in a narrow region of space near the line connecting the quarks.

(...) Mechanical image, which leads to the dual theory is the idealization of the quark-parton chain should be called not string, the length of which we used to take unchanging, but spring. It will be shown that this object is similar to that "spring", the American toy "a slinky", which is used in the Krauford's textbook for imaging wave phenomena. This thin elastic helix wire consisting of a large number of turns.

(...) A strange thing happens in the attraction between two quarks: the strong force does not decrease with the distance between the two particles, as the electromagnetic force does; in fact, it increases, more akin to stretching a mechanical spring.

(...) The interaction of the particles and resonances is depicted by the breaking or association of the strings. (...) When the strings break, it occurs a partial compression of them, and then the elastic energy passes into the flight kinetic energy. This picture corresponds to the cascade mechanism of decaying the resonances."

As anyone can see, the classical model of spinning ring electron gives a picture of the strong interaction largely similar to that resulted from experimental research, but in which the quarks stop being mysterious objects and become entities easy to be understood as respects their origin, the strength of their mutual interactions and other main characteristics of them. Also, as their existence is possible strictly inside a space with a huge magnetic induction, maybe up to 10^{17} ... 10^{18} T, out of which their metamorphosis is practically instantaneous in our scale of time, it becomes quite understandable why the quarks cannot be noticed in free state.

Before ending a last remark: all hadron structures are unstable, except that one existing in the center of the proton as a small but hard core, which contains almost the whole mass of the proton, but has null kinetic and magnetic moments owing to the equal number of its constituent coaxial quarks and antiquarks. Therefore, as the whole dipole magnetic field of the proton belongs to its external positive ring meson, that more recent opinion according to which the nuclear forces could be a residue of the strong forces acting inside the nucleons is false.

Two very important conclusions emerge clearly from the two previous sections:

(1) Magnetic nature of both weak and strong interactions means evidently a decisive step in the direction of the much desired unification of the first three interactions identified until now as being fundamental in nature: weak, electromagnetic (somewhat improperly called so, as long as its non-dimensional constant $g_{em} \approx 1/137$ on the scale where $g_s \approx 1$ is equal to the non-dimensional fine-structure constant $\alpha = e^2/4\pi\varepsilon_0 \hbar c$, and this fine structure of spectral lines is a purely magnetic effect) and strong. Initially a partial unification was argued for electromagnetic and weak interactions[99], but later the researchers found more and more indicia for adding to them the strong interactions, in a *great* unification always predicted but never carried through until now.

And this failure is quite understandable, as long as in quantum relativistic physics all the electrons, mesons and quarks are considered pointlike particles with no inner structure. On the contrary, the long dreamed unification of weak, electromagnetic and strong interactions results by itself when that nonsensical concept of pointlike elementary particle accepted by need in quantum relativistic physics is abandoned and the inner structures of all elementary particles are deduced starting from the Heisenberg's conclusion about an universal matter of which these particles are.

More, this classical unification includes naturally the nuclear interactions whose origin is still uncertain today, as long as the effective action radius of the strong forces existing inside the hadrons are exceedingly small

---

[99] S. Weinberg, *Unified Theories of Elementary-Particle Interaction*, Sc. Amer. **231** (1), (1974) 50.

as against the known sizes of the two nucleons and the known distances between them in nuclei, so that these strong interactions acting inside the nucleons cannot have even the slightest contribution to the nuclear interactions in nuclei, as claiming the latest theories in the field, after several others have been abandoned in the last half century.

(2) If all mesons, either free or bound inside the hadrons, as well as the quarks are in fact spinning ring electrons contracted in very strong magnetic fields, in a still incompletely known chain of successively reversible transformations $e^{\pm} \rightleftarrows \mu^{\pm} \rightleftarrows \pi^{\pm} ... \rightleftarrows q^{\pm}$ (or q and $\bar{q}$), and the photons and neutrinos are only inert emanations of the spinning ring particles, spatial structures made of preons much too far from each other to interact between them and consequently destitute of any internal cohesion, structures always radiated by the ring particles for conserving in any situation their immutable spin $L_e = 1.0545 \cdot 10^{-34}$ J·s, then the ring electrons and positrons remain the only elementary particles in the near microcosm. Therefore, the classical model of spinning ring electron is in harmony with the older hypothesis that all elementary particles in microcosm comes from electrons and positrons, assumed by many physicists for a long time[100].

Gravitational interaction

When the stalk of an apple breaks, that apple falls to the ground because it is attracted by the Earth, or something pushes it towards the center of the Earth? Well, this question is only apparently naive, actually it calls into question the very nature of gravity.

As we know, several physicists, including Newton, postulated long ago a universal gravitational *attraction* between two bodies and even a law in $1/r^2$ for this attraction, but without a physical ground of it. However, soon enough after that a very simple mechanism of universal attraction was im-

---

[100] E. P. Wigner, C. L. Critchfield and E. Teller, *The Electron-Positron Field Theory of Nuclear Forces*, Phys. Rev. **56**, (1939) 530.

L. de Broglie, *Sur la nomenclature des particules,* C.R. Acad. Sci. **247**, (1958) 1069.

F. Halbwachs, *A new classification of elementary particles and resonances*, Nuovo.Cim. **28**, (1963) 695.

P. F. Browne, *Relativistic States of Positronium and Structure of Fundamental Particles*, Nature 211, (1966) 810.

agined, today totally forgotten, and Feynman[101] explains why it was abandoned, unfortunately through a very superficial and partisan argument:

"Many mechanisms for gravitation have been suggested. It is interesting to consider one of these, which many people have thought of from time to time. At first, one is quite excited and happy when he "discovers" it, but soon finds that it is not correct. It was first discovered about 1750. Suppose there were many particles moving in space at a very high speed in all directions and being only slightly absorbed in going through matter. When they are absorbed, they give an impulse to the Earth. However, since there are as many going one way as another, the impulses all balance. But when the sun is nearby, the particles coming toward the Earth through the Sun are partially absorbed, so fewer of them are coming from the sun than are coming from the other side. Therefore, the Earth feels a net impulse toward the Sun and it does not take one long to see that it is inversely as the square of the distance because of the variation of the solid angle that the Sun subtends as we vary the distance.

What is wrong with that machinery? It involves some new consequences, which are not true. This particular idea has the following trouble: the Earth, in moving around the Sun, would impinge on more particles, which are coming from its forward side than from its hind side (when you run in the rain, the rain in your face is stronger than that on the back of your head!). Therefore, there would be more impulse given the Earth from the front, and the Earth would feel a resistance to motion and would be slowing up in its orbit. One can calculate how long it would take for the Earth to stop as a result of this resistance, and it would not take long enough for the Earth to still be in its orbit, so this mechanism does not work."

Undoubtedly, Feynman knew very well what a circular demonstration is and what value such a twist around the tail has, only so one can explain why he avoids in his pleading an essential *detail*, specified yet in a further section of his lecture:

"According to Newton, the gravitational effect is instantaneous, that is, if we were to move a mass, we would at once feel a new force because of the new position of that mass; by such means we could send signals at infinite speed. Einstein advanced arguments which suggest that we *cannot send sig-*

---

[101] R. P. Feynman, R. B. Leighton and M. Sands, *The Feynman Lectures on Physics*, vol. I, Addison-Wesley Publishing Company. Reading, Massachusetts, 1963.

nals faster than the speed of light, so the law of gravitation must be wrong. By correcting it to take the delays into account, we have a new law, called Einstein's law of gravitation."

Therefore, the Feynman's calculus is based on relativistic postulate limiting the velocity of any interaction, including gravity, to that of the light $c = 3 \cdot 10^8$ m/s. In such a relativistic context, a fall of the Earth to the Sun by decreasing gradually its linear orbital velocity $v = 3 \cdot 10^4$ m/s, owing to the mentioned braking effect, could indeed be taken into account, because the difference between the magnitude orders of the two velocities is not too high. But even in this variant Feynman ought to have had in view that the distances between Sun and planets vary in time, there is even a clear geological evidence for a very slow increase of them, so that any braking effect in the Earth's movement could only be a factor whose role in the whole dynamical equilibrium would be just to compensate constantly a dominant contrary tendency of the Earth for moving away from the Sun.

Still the most evident (and not at all innocent) mistake in the Feynman's reasoning is the velocity of light a priori and illogically assigned to the hypothetical particles presumed to cause the universal *attraction* by their weak absorption in matter (or only by a transfer of momentum during their elastic collisions with the smallest subcomponents of the dense matter). Indeed, if these particles "moving in space at a very high speed in all directions" would have a speed much higher than c, and not at all "infinite", an enormity which is not worth discussing, then the difference between the relative speeds of the particles coming from forward and, respectively, from back becomes entirely negligible, and the corresponding braking effect too.

Moreover, this relativistic limitation of the speed of some hypothetical *gravitons* is not only unacceptable in an argumentation in favor of general relativity, but even nonsensical. If gravitons really exist in a very profound microcosm, only if their dimensions are exceedingly small, and their velocities are exceedingly high, in comparison with what it is noticed in the near microcosm, or the world of the known elementary particles, their penetrability through these elementary particles could be high enough to justify only a very weak absorption of universal gravitons, or only very rare collisions of them with the correspondingly small subparticles existing inside the common matter, but with mechanical effects noticeable at macroscopic level. Or, although conforming to quantum relativistic physics the pointlike photons and neutrinos have the highest velocities in nature, and consequently they must also have the highest penetrability in matter — because what can be more penetrating in dense matter than a pointlikle particle of highest velocity? — in fact their penetrability in dense matter is extremely far from that necessary for gravitons, and this huge discrepancy makes absurd right from the start the Feynman's argumentation.

On the other hand, this idea of a far off microcosm populated by exceedingly tiny particles with huge mass density and velocities would be a natural farther step after that one already done from our macrocosm, populated by large and slow bodies, to the near microcosm populated by the known elementary particles, much smaller, but also much faster and denser. In addition, this hypothesis of a universal isotropic radiation of exceedingly small subparticle in the depths of microcosm, called here gravitons, is not at all something utterly fancy and unheard, it already has acknowledged correspondents even at macroscopic level, as the universal isotropic radiation of microwaves $\lambda = 0.211$ m , or the universal "bath" of neutrinos alleged by quantum relativistic physics.

Beyond doubt, gravitation is a phenomenon entirely comprehensible as a purely mechanical effect, which could occur at a very profound level in a very far microcosm, but with consequences directly noticeable at macroscopic level. Its classical mechanism rejected by Feynman with an unfair argumentation is not only physically possible, but even very plausible.

Another theme of attack against the classical understanding of gravitational interactions is the alleged incapacity of the Newton's law of gravity to explain the so-called *fine relativistic effects* known as experimental tests of general relativity. How true all those referring to the invariant speed of light in gravitational field are, we could see in a previous chapter. However, it is still necessary to approach the problem of these small parts of the whole advances of perihelion of the inner planets in the solar system called *residual* (or *anomalous*, or *in excess*, etc.) because their existence could not be justified any longer by the mutual attractions between the planets, the only gravitational interactions considered by astronomers in the Sun's reference system for justifying the perihelion advances noticed by them to the inner planets Mercury, Venus, Earth and Mars.

First calculated values of these residual advances of perihelion were given by Le Verrier in 1859, who found $38".3/cy$ based on several observations carried out between 1697 and 1848, and Newcombe in 1882, who found 42".95/cy by adding some new perihelion transits of perihelion. And because these residual advances of perihelion were considered by some prominent physicists to be definitively incompatible with Newtonian law of gravity, at the beginning of the past century the whole community of physicists was expecting something new able to solve this deadlock.

These were the circumstances in which Einstein chose this unsolved question as the main target of a more comprehensive version of his special relativity, called general relativity and able to justify the invariant velocity of light even in gravitational field. He began working in this matter in 1907, but his way proved to be unexpectedly long and very hesitating. First results in 1913 were not published any longer, because the new field equations of gravitation predicted a ... regression of Mercury's perihelion. A little later Einstein retracted a paper where its new theory of gravitation succeeded indeed to lead to an extra-Newtonian advance, but of only 18"/cy, and later a manuscript sent to Berlin Academy was three times successively submitted and withdrawn for remaking the field equations until its final version where the calculated extra-Newtonian advance of Mercury's perihelion was, at last!, the desired value 43"/cy [39]. Conforming to his own letters, Einstein was extremely happy at the end of this calvary of successive adjustments, but the troubles reappeared in 1920 when he was charged with plagiarism, because his final equation for extra-Newtonian advance of the Mercury's perihelion proved to be identical with that one derived for the same object by Gerber in 1898. At that moment Einstein had an angry reply, denying have had any knowledge of the Gerber's equation when he made it actual again under his signature, but unfortunately for him even his letters in 1916 were showing that this reply was an untruth. For all that, as Einstein has already been the new guru of physics frequently present in the main newspapers in the world, and after only two years he even got the Nobel prize, all these very unpleasant circumstances were quickly overlooked, and finally the residual advance of Mercury's perihelion was declared a fine relativistic effect which demonstrates the limits of the Newton's law of gravity, and so it has remained until today, despite all subsequent astronomical observations in disagreement with the Gerber-Einstein equation.

Indeed, soon after his wavering start, Einstein was faced for several decades with a disproof of its value 43"/cy, more exactly with the constant growing rate predicted by his theory for *extra-Newtonian* orbital advance of the inner planets in the solar system, the same in all time intervals, no matter what initial and final moments they have. But this rigorous constancy in time predicted by his theory was contradicted by many astronomical observations, which showed a permanent fluctuation of that perihelion advances noticed for the inner planets which could not be explained when only the mutual attractions between them are considered. And these disagreements between the noticed positions of these planets and those predicted by theory were particularly evident even in shorter periods of time, for example for two consecutive transits of Mercury, when sometimes the astronomers found even orbital regressions of the planet as against its expected position! Or, these apparently irregular variations incompatible with Gerber-Einstein equation could be a consequence of an elementary reality in celestial mechanics, but whose approach in specialized literature has always been relentlessly censored: all the planets in the solar system orbit around the mass center c_m of the whole solar system, and not around the Sun's mass center, as it has always been considered simply because all noticed transits of the planets can be related only to a motionless Sun. And when the Sun's referential is replaced by that one of the barycentre c_m it appears little changes in the orbital parameters of the planets, but whose rigorous quantification is exceedingly difficult.

Fortunately the almost insuperable difficulties of a rigorous calculation of the revolution periods of the planets in the Sun's reference frame can be avoided by starting from the simplest system in gravitational equilibrium, that one made of only two celestial bodies with circular orbits and very different masses, $M \gg m$.

In such a gravitational system the astronomers can follow the movement of the small planet m around the massive star M only in the reference frame of the motionless star, wherein the gravitational attraction of the planet on the star is ignored, and for this reason they use the approximate formula $T = 2\pi\sqrt{r^3/GM}$ for the revolution period of the planet, where G is the constant of universal attraction and r is the apparent orbital radius of the planet, equal to the distance between the mass centers of the massive star and the small planet, the latter seen as a material point with negligible gravitational field. Of course, this formula results by equalizing the centripetal $F = GMm/r^2$ and centrifugal $F = mv^2/r$ forces acting on the planet in this referential, where $v = 2\pi r/T$ is the orbital linear velocity of the planet.

But in reality both the star and the planet revolve with the same revolution period around the mass center c_m of the whole binary system, which is placed on the line joining the mass centers of the two constituent bodies in accordance with the lever rule $MR = m(r - R)$, where R is the orbital radius of the massive star and $(r - R)$ is the real orbital radius of the planet. Or, in this reference frame the attraction force exerted by the star on the planet remains the same, but owing to the smaller orbital radius of the planet both the centrifugal force acting on it and its orbital liner velocity change, so that the real period of revolution T' of the planet is a little smaller than the revolution period T considered by astronomers in the reference frame of the motionless star.

Indeed, in the reference frame of the motionless mass center c_m, which corresponds to a more rigorous approach of the two-body problem because here the gravitational attraction exerted by the small planet on the massive star is not ignored any longer, the small planet has a real period T' of revolution calculated by a different formula, $T' = 2\pi\sqrt{r^3(1 - R/r)/GM} \approx T(1 - R/2r)$, since $R \ll r$. Or, the very small time difference $\Delta T = T - T' = TR/2r$ will inevitably be perceived by astronomers as a small advance δ of the planet on its orbit, as against the position expected by them at a certain moment. When expressed in arc seconds (routinely noted "), this orbital advance becomes $\delta = 1.296 \cdot 10^6 \Delta T/T = 1.296 \cdot 10^6 R/2r$ "/T after each revolution period T, where $1.296 \cdot 10^6$ is the number of arc seconds corresponding to a complete revolution. Evidently this orbital advance noticed by astronomers in their reference frame increases constantly in time, so that in a much longer time interval, for example a century (or $3.156 \cdot 10^9 s$), during which the planet carries out $n = 3.156 \cdot 10^9/T$ revolutions, the noticed orbital advance of this planet also increases n times, $\delta = 1.296 \cdot 10^6 nR/2r = 4.090 \cdot 10^{15} R/2rT$ "/cy, a formula whose most compact shape is

$$\delta = k/\sqrt[3]{T^5} \text{ "/cy},$$

where $k = 2.045 \cdot 10^{15} R/2\sqrt[3]{GM}$ is a constant of the given binary system.

Well, as the orbital motions of the planets in the solar system can also be determined by astronomers only in the Sun's reference frame, while in reality the planets orbit around the mass center of the whole solar system, supplementary very small orbital advances of the planets should appear by the same mechanism, and consequently their values for a century should be quantified for all planets by the same law in $1/\sqrt[3]{T^5}$, where T is the orbital period of the considered planet.

Of course, an accurate calculus of these supplementary orbital advances $\delta = k/\sqrt[3]{T^5}$ "/cy of the planets in the solar system is exceedingly difficult, because the very different revolution periods of the eight planets change permanently the solar system configuration, so that the distance R between the mass center of the whole solar system and the Sun's mass center varies continuously, in very large limits and apparently in a chaotic manner, which changes the constant k into a variable whose mean value for a century can probably be found only by means of some powerful computers.

Still the formula $\delta = k/\sqrt[3]{T^5}$ "/cy can be applied to the solar system for calculating relative values for the supplementary orbital advances of the planets due to their orbital motions around the mass center of the whole system, and not around a motionless Sun, when the constant k disappears through simplification.

Concretely, as the four inner planets Mercury, Venus, Earth and Mars have experimentally measured periods of revolution $T_{Me} = 7.603 \cdot 10^6$ s, $T_V = 1.944 \cdot 10^7$ s, $T_E = 3.154 \cdot 10^7$ s and $T_{Ma} = 5.936 \cdot 10^7$ s, and any time the constant k is the same for all of them, their supplementary Newtonian advances on orbit δ_{Me}, δ_V, δ_E and δ_{Ma} have always to be in determined ratios, valid for any time interval regardless of its initial and final moments, and irrespective of the corresponding absolute values of the perihelion advances noticed to them in a time interval or other:

$$\delta_{Me} = 4.78\, \delta_V = 10.71\, \delta_E = 30.73\, \delta_{Ma}.$$

Or, these relative values of the supplementary Newtonian advances on orbit are very similar to their homologous values resulted from the current relativistic perihelion advances of the inner planets, $\delta_{Me} = 43.03$ "/cy, $\delta_V = 8.6$ "/cy, $\delta_E = 3.80$ "/cy and $\delta_{Ma} = 1.35$ "/cy [102],

$$\delta_{Me} = 5.00\, \delta_V = 11.32\, \delta_E = 31.87\, \delta_{Ma}.$$

Two conclusions are evident:

(1) Since all residual advances of perihelion noticed to the inner planets in the solar system are correctly quantified by a law in $1/\sqrt[3]{T^5}$ proper

---

[102] G. E. Smith, *Closing the Loop: Testing Newtonian Gravity, Then and Now*, in *Newton and Empiricism*, eds. Z. Biener and E. Schliesser (Oxford University Press) 2014.

to Newtonian mechanics, the extra-Newtonian origin of these residual advances of perihelion is clearly invalidated;

(2) Even if the classical formula $\delta = k/\sqrt[3]{T^5}$ "/cy can calculate in the first instance only relative values, and this for evidently objective reasons, the real cause of the so-called *residual* advances of perihelion found by astronomers in their reference frame related to a motionless Sun is as clear as possible. More, this classical solution of the problem is much closer to experimental reality just because it predicts larger or smaller variations of these residual advances of perihelion, depending on the considered interval of time. For example, during a planetary aligning in the solar system, a simple calculus shows for Mercury a supplementary Newtonian advance of perihelion of about 190 arcsec at a single complete revolution (which corresponds to about 20 °/cy!, therefore at the level noticed to the double stars whose common center of mass is very distant from those of the two constituent stars).

Resuming, there is no valid theoretical argument against a classical understanding of gravity, as a phenomenon that occurs somewhere in the depths of the physical microcosm, and also no experimental test of general relativity unsolvable for Newtonian law of universal attraction, including the allegedly extra-Newtonian advances of perihelion noticed to the inner planets in the solar system. In fact, general relativity was invented by Einstein only for saving his postulate on the invariance of the velocity of light from a clear and definitive invalidation in gravitational field.

On the other hand, if the universal attraction can be, excluding a general relativity with no experimental test inexplicable for Newtonian gravitation, just a mechanical effect noticeable at macroscopic level, but occurring in fact in an extremely deep microcosm, anyway beyond that of the preons also subjected to gravitational interactions, a great unification of all the four fundamental interactions in nature, today often called *theory of everything*, remains only another chimera of quantum relativistic physics.

RING ELECTRON WITHOUT ELECTRIC CHARGE

Magnetic but neutral ring electron

By reducing the Planck's constant to a value $h = 3.313 \cdot 10^{-34}$ J·s twice smaller than that one considered until now, which implicitly halves all values $E = h\nu$ placed on the energetic scale of electromagnetic radiation, the spinning ring electron with rest energy $E_0 = m_0 c^2/2$ becomes automatically the only possibility of understanding this fundamental elementary particle. And this classical model of electron proves indeed to be very effective in explaining all the known properties of this particle, as well as for finding the internal structure of all the other elementary particles, structural models in their turn able to justify the main properties of these particles, many of them definitively unexplainable in quantum physics.

But precisely these incontestable arguments make very annoying that evident heel of Achilles of the spinning ring electron, namely the insoluble contradiction between the dipole magnetic field with axial symmetry generated by this spinning ring electron and the pointlike electricity quantum e attached automatically to the electron when this proved experimentally to be not only an indefinite quantum of all macroscopic quantities of electricity, as it has always been considered before 1997, but also an elementary particle with determinate mass. Indeed, on the one hand any dipole magnetic field with axial symmetry can be generated only by a circular electric current, and consequently the elementary electric charge carried by the electron should have a circular form and rotate around its symmetry axis, but, on the other hand, an elementary electric charge with spatial extension cannot physically exist owing to the generalized repulsive forces acting inevitably between all its constituent parts.

Although at present such a fundamental contradiction is evidently unsolvable, an always overlooked suggestion comes yet from the celestial bodies in rotation — planets, stars, galaxies, swarms of galaxies — since all of them generate dipole magnetic fields with axial symmetry even if their matter is always electrically neutral on aggregate, with equally numerous positive protons and negative electrons homogenously distributed inside them, and with no possibility for a spatial segregation of them in distinct zones differently charged. More, the strength of their magnetic fields of rotation depends only on their mass, dimension and angular velocity of rotation. For example, in our solar system Venus and Earth have about the same densities and dimensions, but Venus has a magnetic field much weaker in agreement with its much slower rotation. On the contrary, Jupiter registers on its surface a magnetic induction eight times higher than that on terrestrial surface solely because Jupiter has a rotational angular velocity almost 2.5 times larger than Earth (but this means for Jupiter a peripheral linear velocity at equator 21 times larger than that one of the Earth).

Moreover, the so-called neutron stars, which have huge mass densities estimated to about 10^{14} kg/m^3 and rotation periods measured in seconds, but are not electrically charged, generate yet the strongest magnetic fields of rotation noticed in Universe. And in microcosm the magnetic fields generated by the orbital motions of nuclear nucleons, particles electrically neutral, have been discovered in nuclear spectra through the spectral splits caused by them.

A more complex case in this regard is the Sun. Although magnetic induction measured at its surface is generally similar to that one measured at the Earth's surface, the huge convection currents of solar plasma involved in the solar explosions cause around them temporary increases in local magnetic induction, even several thousand times. Or, these convection currents contain the same plasma existent in the whole star, with a high degree of thermal disorder and a predominance of excited and ionized atoms, in truth, but yet electrically neutral on aggregate, with an equal number of negative and positive electric charges in any macroscopic unit of volume. Therefore, these marked increases in magnetic induction around the convection currents preponderantly perpendicular to the solar surface prove that the electrically neutral solar plasma generates magnetic fields not only by its rotation around an axis, but also by its fast rectilinear motion.

Well, all these well-known realities have an undeniable conclusion: all magnetic fields are generated by the circular or linear movement of the neutral mass, and not at all by the movement of the the electric charges attached to elementary particles, because if the electric charges in motion would be the real sources of magnetic fields, all macroscopic bodies in rotational or linear motion could not give rise to magnetic fields around them, just because their negative and positive electric charges are equally numerous, and consequently their resultant magnetic fields would be exactly null.

But even if we really have clear reasons to consider the dipole magnetic fields of the electrons as being generated by the cyclotron motion of their neutral constituent preons, and consequently to eliminate from this viewpoint the intricate quantum of electricity assigned to any electron, the structural model of spinning ring electron still cannot get rid of this inconvenient question without clearing up the problem of the electric field with spherical symmetry postulated to be generated by any elementary electric charge, either in motion or at rest, a concept deep-rooted in particle physics since its inception.

Electric or magnetic fields?

The concept of electric charge was adopted in the 18th century when a series of experimental observations suggested a granular structure of the electric currents seen until then as continuous fluids, and after Coulomb established empirically the law in $1/r^2$ for the repulsive forces between two little balls charged with negative static electricity, this result could be justified at that time just by assigning an electric field with spherical symmetry, therefore with the same law in $1/r^2$, to each of all the elementary electric charges uniformly spread on the spherical surface of the little charged balls, since only so the latter could have on aggregate macroscopic electric fields also with spherical symmetry and radial force lines, which correspond to the law in $1/r^2$ of their strength.

But after discovering the dipole magnetic field with axial symmetry of the electron the Coulomb's law in $1/r^2$ could be explained in a completely different way, which makes again actual the fundamental dilemma of early electromagnetism: are really the electric and magnetic fields of different nature, or not?

As it is known, in the age before discovering the real nature of electricity quanta, after many discussions pros and cons the final verdict to the previous question was affirmative, in the main because the force lines of the electric fields always issue from some electric charges and end in other of opposite sign, or go to infinity, while the force lines of the magnetic fields are closed curves around electric currents, or go to infinity. And the ultimate argument was the alleged impossibility to obtain a magnetic field with spherical symmetry through any combination of currents or permanent magnets. Unfortunately, for the whole subsequent development of electromagnetism, this assertion has been entirely false. Really amazing, but the clearest example in this regard is just the little charged balls used by Coulomb in his experiments.

Indeed, as at macroscopic level a static negative charge consists in a very big number of motionless free electrons evenly spread on the surface of a body, these free electrons can remain at rest and equally distanced from each other only if all magnetic forces acting on them allow a static equilibrium by their mutual annulment. And such a global static equilibrium is possible, for evident reasons, only if the symmetry axes Oz of the dipole magnetic fields of all neighboring free electrons are parallel between them and perpendicular to the local surface. In this arrangement each free electron exerts repulsive magnetic forces on its neighboring, which on the one hand prevent the free electrons from leaving the surface, and on the other hand assure equal distances between them, therefore a uniform distribution of the free electrons on the surface.

Once reminded these elementary truths, it is simple to shape the magnetic force lines springing from a surface uniformly charged with negative static electricity: because all dipole magnetic fields of the free electrons at rest on a plane portion of the surface have their Oxy planes of symmetry parallel to this surface and magnetic moments perpendicular to it, all magnetic force lines arising to the outside from the equidistant free electrons at rest on a larger surface are mutually forced to become normal to the local surface, no matter what relief the whole surface has, just because the force lines of two coplanar dipole magnetic fields interacting repulsively tend to become parallel to infinity. In consequence, any macroscopic body covered by equidistant free electrons at rest has certainly a macroscopic magnetic field whose force lines tend everywhere to be perpendicular to its surface at a distance from surface much larger than the distances between the neighboring free electrons.

If so, all macroscopic magnetic fields resulted as the vector sum of all dipole magnetic fields generated by the free electrons in static equilibrium on the surface of negatively charged bodies can easy be shaped regardless of the geometrical form of the latter: always their force lines are perpendicular to the local surface (of course, excepting those arising from the marginal parts of some finite charged surfaces, for example a plate, whose their mutual alignment to directions normal on the surface diminishes gradually to the margins, where the magnetic force lines become more and more curved outwards).

Applied to negatively charged spheres, exactly the ones used by Coulomb in his experiments, this general rule leads to macroscopic magnetic fields with radial force lines and implicitly with a law in $1/r^2$ of their strength. Or, the little charged balls used by Coulomb cannot have simultaneously two different fields with spherical symmetry, one electric and other magnetic, and if one of them has to be eliminated, that one can only be the electric field, because the own dipole magnetic field of the electrons is an experimental certainty, while the elementary electric charge and its electric field with spherical symmetry were invented long before discovering the real nature of the electrons and their own dipole magnetic field, by which they can interact remotely with other magnetic particles, both attractively and repulsively. Finally, if the charged ball has a macroscopic magnetic field with radial force lines and a law in $1/r^2$ of its strength, and not a macroscopic electric field with the same characteristics, the free electrons on its surface have no electric charge.

As normal, such an abrupt change in physics has to be confirmed by experimental arguments with unquestionable significances. Undoubtedly this radial magnetic field of a sphere charged with negative static electricity could be directly attested with a very fine magnetic needle protected against electric discharges – and it is even hard to understand why such a direct verification has never been done —but many other indicia in this respect still wait for understanding their significance.

For instance, when the macroscopic field of a very tiny spherical body covered with negative static electricity (now improperly called *pointlike* charge of electricity) is visualized by means of some little acicular particles, as little splinters of gypsum initially chaotically spread on a paper, or of some oblong seeds floating in an insulating liquid, their alignment head-to-tail along the radial force lines of the field can be a clear indication about the real nature of this field. Unfortunately, as this alignment to the field lines involves evidently a rotation of the oblong particles around an axis passing through their mass center, and this rotation can be determined only by a couple of antiparallel forces acting simultaneously on their two ends, for more than a century this turning on the spot of the oblong particles has been wrongly explained by the action of the external radial electrostatic field on the two differently charged ends of the oblong particles, became electric dipoles in this field owing to the internal migration of a certain number of weakly bound electrons along the field lines. But such a different electric polarization of the two ends of all oblong particles randomly orientated in an electrostatic field is a premise definitely absurd, because such an electric polarization could occur strictly along the radial electric force lines which intersect the oblong grains, and these lines pass very seldom through both ends of the oblong grains chaotically oriented to one another (excepting, of course, those oblong grains accidentally already aligned to the local force lines of the field), so that the two charges induced within the grains cannot evidently give birth a torque able to align these oblong grains to the local flux lines. It is therefore beyond doubt, such electric polarizations strictly along the radial directions which pass through the randomly orientated acicular particles are totally unfit for originating couples of forces able to align that acicular grains to the local force lines. On the contrary, these electric polarizations of the splinters inside the external electric field with radial force lines could at most induce some weak attraction forces exerted on them strictly on the radial directions of the field, towards the central macroscopic electric charge.

It is also very clear, an alignment by rotation of the oblong particles along the force lines is the typical behavior of a magnetic needle in an external magnetic field. And after such a rotation two parallel oblong particles reject each other as two parallel permanent magnets with the same orientation $N - S$, so that they set on equidistant radial force lines, in a head-to-tail succession on each force line.

Another at hand way for discerning between electric or magnetic nature of the fields generated in surrounding space by negatively charged bodies at rest is a more attentive examination of the shape of their force lines in regions beyond their edges, for instance in the case of a thin plate covered with negative static electricity on both faces. Pictures of this kind can be found in many textbooks or treatises, but unfortunately their fields of sight end always too close to the edge of the charged plate (as for example in[103]). Or, the shape of these force lines becomes significant only at a certain distance from the charged plate: indeed, if the visualized force lines belong to a macroscopic electric field, they have to have trajectories constantly more distanced from the plane of the plate, but if the macroscopic field of the charged plate is of magnetic nature, at certain distances from its margin the force lines have to curve towards the plane of the plate before going to infinity on directions more and more parallel to this plane. Evidently such a curvature towards the plane of the electrically charged plate could not be justified by electric force lines, whose paths have to move endlessly away from the charged plate from which they spring.

The force lines joining the two parallel plates of an electric capacitor, one covered by free electrons uniformly spread on its surface and the other by a layer of marginal atoms with a deficit of electrons, have a similar shape, normal to the surface just in the central region of the plate, and more and more curved outwards at their edges. These lines are considered to be electric field lines, but they are identical with the magnetic force lines between the near N and S poles of two permanent magnets with the same section and collinear axes of symmetry.

More, Coulomb himself found in his experiments a force in $1/r^2$ acting between the two near poles of two collinear long permanent magnets, provided the distance between their two poles to be small in comparison with the length of the two magnets, but large in comparison with their transversal section.

For all that, neither Coulomb and nor someone after him verified if the radial field of a little ball charged with negative electricity is not also of magnetic nature. And if in the Coulomb's time the dipole magnetic fields of electricity quanta were not known yet, this lack of curiosity in the modern age is hard to understand.

Actually all macroscopic bodies negatively charged generate around them magnetic fields whose force lines have a shape which can be exactly drawn by summing all dipole magnetic fields generated together by the free electrons on their surface, whose magnetic moments are always normal to the local surface and parallel between them. And the best example remains

---

[103] D. Halliday and R. Resnick, *Physics*, Part II, Wiley & Sons, New York, London, Sidney, 1966, Fig. 27-6 (a).

the macroscopic magnetic field of a negatively charged sphere, with radial force lines and consequently with a law in $1/r^2$ of its strength.

Moreover, besides the existence of the elementary electric charges solely to magnetic elementary particles (the neutral but magnetic neutron does not infirm this rule, as long as its magnetic moment is in fact the vector sum of two collinear magnetic moments which belong to a positive proton and a negative meson), these enigmatical elementary electric charges raise justifiably another serious queries concerning their apparition during any creation of pairs electron-positron, $\gamma \to e^- + e^+$, when the summed mass of the two new formed electrons is rigorously equal to the mass of initial photon, but the two electric charges attached to them appear from nothingness, and also concerning their disappearance with no trace during an annihilation process $e^- + e^+ \to 2\gamma$, when each of the two new formed photons takes integrally the mass of the electron out of which originates, but both electric charges attached to initial electrons disappear like ghosts. Or, in a concrete material world, the universal law of conservation cannot be primitively reduced to an arithmetical summing $(+1) + (-1) = 0$, all the more so as the concept of positive and negative electric charges introduced by Franklin is only a convention with no physical content. As a matter of fact, after some centuries from introducing this still so mysterious concept of electric charge and despite the huge development of experimental research in physics of the modern age, besides the absolutely conventional partition of the elementary electric charges in positive and negative quanta, nothing else can be said about their physical nature, so that their existence increasingly seems to be rather imaginary than real.

Forces acting on electrons in magnetic fields

There are three distinct kinds of action exerted by the external magnetic fields on the electron. Two of them are identical with the actions exerted at macroscopic level on the circular electric currents, but they cannot be explained if the electron is a pointlike, non-dimensional body, while the third cannot be explained neither by the ring electron model and the less by the pointlike electron with no inner structure.

(1) When the electron enters a non-uniform magnetic field, it is accelerated along the non-parallel force lines of these fields. This effect has been

used in Stern-Gerlach equipments for estimating the magnetic moments of paramagnetic particles by measuring their deviation in non-uniform magnetic fields perpendicular to their motion direction. The force acting on the electron is $F_e = L_e \cdot \Delta B / \Delta d$, where the non-uniformity of the magnetic field is given as a ratio between its induction variation ΔB along a very small distance Δd, and this formula is entirely similar with that one deduced for macroscopic circular currents, which confirms once again the similitude between the electron and an infinitesimal circular electric current. On the contrary, the pointlike electron of quantum relativistic physics is definitely incapable of justifying its acceleration in a non-uniform magnetic field, whereas the non-uniformity of a magnetic field can be detected only by comparing magnetic inductions existing in at least two different points of the field, something evidently impossible for a pointlike entity.

(2) The electron behaves also similarly to a macroscopic circular current by the permanent tangential alignment of its magnetic moment to the local force lines of all external magnetic fields, either uniform or non-uniform. Evidently such an alignment involves a tipping of its magnetic moment, either at the entry into a uniform magnetic field, excepting the case when its magnetic moment is by chance already parallel to the respective field lines even before entering, or permanently in a non-uniform magnetic field, and such a tipping can evidently be determined only by the action of a couple of antiparallel forces. Or, if such an alignment of the electron magnetic moment to the field direction is easy to understand in the case of the ring electron, it becomes quite incomprehensible for the pointlike electron, as long as any spatial overturn of a body needs a couple of forces with two different points of application.

(3) A still intricate action on the moving electron is that noticed in a plane perpendicular to a uniform magnetic field $\bar{B}$, where any electron performs endlessly a uniformly circular motion called cyclotron motion. This motion was always explained by the action of the external uniform magnetic field through a specific force $\bar{F} = e\bar{v} \times \bar{B}$, called Lorentz (or Laplace) force, which is permanently both coplanar with the linear velocity vector $\bar{v}$ of the electron and perpendicular to it, like the centripetal force in classical mechanics. However, although the above formula of the Lorentz force describes correctly the cyclotron motion, at least for the electrons with relatively small velocities $v \ll c$, this understanding of cyclotron motion infringes clearly the universal law of angular momentum conservation: indeed, when an elec-

tron with mass m and linear velocity v enters a uniform magnetic field B on a direction perpendicular to the field direction and begins an uniformly circular motion with orbital radius r, the new orbital angular momentum $L_{orb} = mvr$ cannot appear in the isolated system electron-magnetic field without a corresponding variation of other angular momentum in this system, just because only so the universal law of angular momentum conservation can be observed. Another explanation simply cannot exist, and perhaps the best illustration of this elementary truth is that well known example of angular momentum conservation used in many treatises of mechanics, with a man that stands on a rotating platform at rest and holds overhead the vertical axis of a wheel in very fast rotation.

Indeed, when this man inclines the wheel axis to horizontal, the rotation speed of the wheel remains the same, but simultaneously the whole platform begins rotating on its own central axis with the angular velocity necessary for preserving unchanged the total angular momentum on the vertical direction in this isolated system man-platform, although apparently no force acts on the platform. And the platform initially at rest begins rotating even when the axis of the spinning wheel remains vertical, but the man changes, one way or other, the rotation velocity of the wheel. Therefore, any change in the angular momentum of the spinning wheel, either as orientation or as absolute value, triggers *automatically* the rotation of the entire platform, as the only possibility for observing the angular momentum conservation on the vertical direction. Obviously the cyclotron motion of an electron in a plane perpendicular to a uniform magnetic field can appear only by a mechanism similar in principle, wherein the universal law of angular momentum conservation is rigorously observed.

Therefore, when an electron with linear velocity $\bar{v}$ enter a uniform magnetic field $\bar{B}$ on a direction perpendicular to the field lines, $\bar{v} \perp \bar{B}$, and simultaneously it begins a cyclotron motion, which can only appear for compensating another angular momentum variation determined by its entering the field, this first angular momentum variation can imply only the electron spin, more exactly only its spatial reorientation by aligning to the field lines, whereas the absolute value of the electron spin $L_e = m_0 c r_e$ remains always the same, irrespective of any variation of the external magnetic induction.

For example, when one β-electron or electron previously electromagnetically accelerated enters a uniform magnetic field $\bar{B}$ with linear velocity

$\bar{v}$ perpedicular to the field lines, its longitudinal spin polarization $\bar{L}_e \perp \bar{B}$ becomes instantaneously transversal $\bar{L}_e \uparrow\uparrow \bar{B}$, and this tipping of 90° is an angular momentum variation whose compensation (obligatory conforming to the universal law of angular momentum conservation in isolated systems) can only be an orbital angular momentum in the only plane where its linear velocity has a certain value $v \neq 0$, which evidently is that one perpendicular to the field lines. And this induced orbital angular momentum $L_{orb} = mvr$ has to be equal to the electron spin $L_{orb} = L_e = m_0 c r_e$, which is easy to demonstrate through a very simple vector calculus.

Curiously enough, experimental data able to confirm or infirm this logical understanding of cyclotron motion, actually the only in accordance with the law of angular momentum conservation, cannot be found in literature despite the high stake of this question, since an eventual invalidation of the concept of Lorentz force means at the same time an invalidation of the elementary electric charge assigned to the electrons, at least as entities upon which the external magnetic fields actually act. The clearest conclusions in this regards could be obtained by specific experiments using electrons with transversal spin polarization, whose behavior after entering uniform magnetic fields on directions perpendicular to the field lines would clarify everything, and the synchrotrons show how simple is the change of the electron spin polarization from longitudinal in transversal.

Attractive interactions between electrons

As all magnetic dipoles, the electrons can have magnetic interactions both attractive and repulsive, depending both on their mutual positioning in space and the mutual orientation of their magnetic moments. But if the electrons have also quanta of negative electricity, whose interactions are exclusively repulsive, they can stay together in very small spaces and form there stable configurations only if their magnetic interactions are attractive and a little stronger than the electric repulsions which tend to disperse them. Concretely, as the permanent electric repulsion between two electrons is given by the formula $F_e = e^2/4\pi\varepsilon_0 d^2$, where $\varepsilon_0 = 1/(36\pi \cdot 10^9)$ F/m is permittivity of vacuum and d is the distance between them, while the possibly attractive magnetic force between them can be estimated by the formula $F_m = \mu_0 M_e^2/4\pi d^4$, where $\mu_0 = 4\pi \cdot 10^{-7}$ H/m and $M_e = 9.285 \cdot 10^{-24}$ A·m², a simple calculation proves that $F_m > F_e$ is possible only when the distance d between the two electrons decreases to $(2...3) \cdot 10^{-13}$ m (depending on the mutual orientation of their magnetic moments).

But, if so, some common situations become hard to understand, as for example how can the free electrons stay at rest on the surface of a body charged with static negative electricity, as long as each of them is subjected to all electric repulsion forces generated not only by the neighboring free electrons on the surface, but also by the nearest marginal electrons in the atomic lattice of the body, which tend together to send any free electron in the surrounding space, and the only possible attractive force acting against these repulsive forces could result by its magnetic coupling at a distance smaller than $3 \cdot 10^{-13}$ m with one marginal atomic electron in the atomic lattice of the charged body, a distance at which a very slow electron cannot anyway reach owing to all repulsive forces mentioned before, much stronger at distances larger than $3 \cdot 10^{-13}$ m .

Atomic lattices moot the same question. Two or several atoms with central positive nuclei about 10^4 times smaller than their marginal layers of negative electrons, cannot bring together as long as, most evidently in the case of heavy atoms, their few unpaired peripheral electrons cannot compensate through their attractive magnetic pairing the much stronger electric repulsions between all peripheral electrons existing at each moment in the touching spaces of the two atoms. And as for that sometimes invoked electric attractions exerted by atomic nuclei on peripheral electrons in the surrounding atoms, their effect is clearly quite negligible in comparison with the general electric repulsion between all atomic electrons existing in the space between the two nuclei, just because electric interaction has a law in $1/r^2$ and the mean distances between the electrons present in the spaces between nuclei is many times smaller than atomic radii.

Similar questions arise about all covalent links between identical or different atoms, which are always made by the same magnetic pairing between the valence electrons initially unpaired within the free atoms, but in this case the question is not only how can the involved atoms approach near enough for an attractive magnetic coupling of their valence electrons, but even how their pairs of pooled electrons can coexist in the small interatomic space wherein the covalent links are placed. For example, how can the six electrons of a triple covalent link $N \equiv N$ coexist in the small space between the two bound N atoms, as long as each valence electron is attractively coupled by magnetic interaction with only one valence electron in the other atom, but it is simultaneously rejected through electric interactions by all the other five valence electrons in its proximity? And again the answer is clear: such a compact gathering of atomic electrons *put in common* by the two N atoms can exist only if all electric repulsions between all the six valence electrons simply do not exist.

Resuming, all stable and compact agglomerations of electrons, whether on the surface of some macroscopic bodies, or between the atoms condensed in atomic lattices, or between the atoms covalently bound by attractive magnetic interaction, and so on, could not appear if their constituent electrons would really bear a negative elementary electric charge, just because in this case their mutual electric repulsions would make impossible their closeness at distances small enough for magnetic pairing. And yet, such stable and compact groups of very close electrons exist in different variants.

Atoms made of nuclei and electrons without electric charges

The critical review of the main functions attributed to elementary electric charge attached to the electron reveals conclusively their falsehood, so that the spinning ring electron destitute of such an elementary electric charge becomes a structural model not only with no conceptual contradiction, but also able to justify both the own dipole magnetic field of the particle and the Biot-Savart magnetic fields generated by its motion, the spatial shape of the allegedly electric fields generated by one or more electrons at rest, which after a more attentive examination prove to be in fact magnetic fields, all interactions with other elementary particles, including attractive interactions with other electrons, either in atomic lattices or in molecules, why and how the velocity and the trajectory of the electron change in external uniform or non-uniform magnetic fields, and so on.

In this context, since all atomic theories, from Bohr's semi-quantum theory to modern quantum atomic physics, are clearly disproved by many experimental data and before anything else by the measured hyperfine and fine splitting of the energy levels in hydrogen spectrum, they become a very important argument in favor of the neutral spinning ring electron, just because all these atomic theories start from the postulated attraction between the positive and negative quanta of electricity carried by protons and, respectively, the electrons. Logically, the proved falsehood of these atomic theories means implicitly the falsehood of their premises, including the forming of the atoms through electric attractions between positively charged nuclei and negatively charged electrons.

However, if these electric attractions between the positive protons and the negative electrons do not exist simply because the elementary electric charges attached to these particles are not real, how could the neutral nuclei bind around them a determinate number of electrons with constant linear velocities, proved by the determined values, always the same, of the energies necessary for removing from atoms their different electrons, and on strictly circular orbits with constant radii, proved by the determined values, always the same, of the hyperfine and fine splits of all energetic levels in atoms?

Any attempt to find a reliable answer to this very complex and difficult question, insoluble if all atomic electrons interact repulsively between them owing to their identical electric charges, has evidently to take into account only intraatomic magnetic interactions, attractive or repulsive depending on the spatial positioning of the atomic electrons and the mutual orientations of their magnetic moments: those attractive responsible for the pairs of atomic electrons with antiparallel magnetic moments, reason for which the atoms have total magnetic moments never higher than some few Bohr magnetons, and those repulsive responsible for the symmetrical positioning of the atomic electrons on their atomic orbits, as well as for the limited number of electrons on each atomic orbit.

This idea of stable planetary systems made of n-body interacting through their magnetic fields was applied first time to the solar system (Gilbert, 1600), but a century ago Parson[24] used it for explaining the determinate number of electrons on the successive atomic orbits and the total magnetic moments of the electrons, and in the next decades Frenkel, Tamm, Franck, Ginzburg or Heisenberg studied different variants of such magnetic configurations in dynamic equilibrium and their applicability to nuclear structures with determined kinetic and magnetic moments. Significantly, among all kinds of permanent magnets studied by them to this end, the superconducting rings of current proved to be the most suitable, since they can form stable planetary systems irrespective of the perpendicular or parallel spatial orientation of their magnetic axes as against the orbital plane.

But perhaps the most significant contribution to idea of atoms shaped exclusively through magnetic interactions remains an older experiment[104] with vertical magnets floating in a liquid in a vessel with circular section, wherein they had evidently the same vertical direction of their moments. Initially these vertical magnets representing *atomic electrons* float toward the rim because of their mutual magnetic rejection, but when an electromagnet is placed under the vessel, then all these *electrons* are grouped in a series of concentric rings (*orbits*) whose number is determined by the number of *electrons*. What better incentive is necessary any longer for researching in depth the version of magnetic atoms?

Two very difficult questions appear yet:

(1) How can an initially free electron become a satellite of a nucleus?

(2) How can all atomic electrons have orbital motions with constant linear velocity, the only able to justify the exactly determined values of all ionization energies in atoms, and also with constant orbital radii, the only able to justify the exactly determined values of all hyperfine splits of the energy levels in atomic spectra?

The answer to these questions unsolvable for the Bohr's atom must be found starting from the quantizing equation $mvr = \hbar = 1.0545 \cdot 10^{-34}$ J·s valid for both 1s and 2s states in hydrogen and deuterium atoms conforming to experimentally measured hyperfine splits of their 1S and 2S terms, which differs essentially from the Bohr's postulate $mvr = n\hbar$ taken in all atomic theories, including the modern quantum atomic theory. Or, such an identical orbital angular momentum even only for both 1s and 2s states of these atoms means automatically the same orbital angular momentum $mvr = \hbar$ for all atomic electrons in all atoms, because the same orbital angular momentum in two different atomic states of the same atoms can only be just the consequence of a unique modality of binding the free electrons to nucleus, one and the same for all similar processes.

Indeed, when a free slow electron is magnetically attracted by a much heavier nucleus, it goes not along the straight line joining them, but always

---

[104] V. V. Kozoriz, *Dynamic Systems of Magnetically Interacting Free Bodies*, Naukova Dumka, Kyiv, 1981 (in Russian).

along the local force line of nuclear magnetic field, and this route ever curved means implicitly a continuous change of the spatial orientation of its spin, every moment obliged to be tangent to that curved force line. Or, conforming to the universal law of angular momentum conservation, any change of spatial orientation of a vector quantity, as the electron spin is, has to be compensated by other angular momentum variation of equal amplitude, in this case by an orbital angular momentum of the considered electron, so that the initial total angular momentum to be preserved on any spatial direction, somewhat similar to the angular momentum conservation in that previous classical exemplification with a man holding a spinning wheel in his hand on a rotating platform.

And if all electrons bound in atoms in this way specific to magnetic interactions between magnetic dipoles change the spatial orientation of their spin with an angle equal to $\pi/2$, for example if initially they have a longitudinal spin polarization and in the final bound state a transverse spin polarization, all of them should have orbital motions around nuclei, always with constant linear velocity and orbital radius, always with the same orbital angular momentum equal to the electron spin, $L_{orb} = L_e$. Obviously, such perfectly uniform and circular motions of the atomic electrons around atomic nuclei are possible only if they are induced by the universal law of angular momentum conservation in an isolated system, while in the atoms made of nuclei and electrons with elementary electric charges such perfectly uniform and circular motions of the electrons are absolutely impossible owing to the mutual perturbations caused continuously by the repulsive forces between these elementary electric charges. Or, just the determinate values of ionizations energies in atoms, as well as those of the hyperfine splits of their energy levels, require compulsorily constant linear velocities and constant orbital radii in the orbital motions of atomic electrons, therefore their uniformly circular motions within the atoms.

Besides, in the atoms formed exclusively through magnetic interactions the orbital motions of the electrons cease being accelerated motions caused by a central force, these perfectly uniform and circular motions are in fact *inertial motions on circular trajectories*, and this last statement should irritate nobody, because in fact a purely rectilinear motion is possible nowhere in nature, as long as there are no real spaces with absolutely null magnetic induction, always $B > 0$, and consequently always what we consider to be a rectilinear trajectory is in fact only a curved trajectory with very high radius of curvature. In this context the Bohr's postulate about the *stationary* orbits in atoms, where on the atomic electrons do not radiate although their motion is accelerated, becomes completely pointless.

In brief, the spinning ring electron without elementary electric charge, but with a proper dipole magnetic field generated by the subquantum motion of its mass, has not only the big advantage to be exempted from all contradictions tied to the intricate concept of elementary electric charge, either pointlike or with a certain spatial extension, but also a decisive support in a series of experimental arguments in its favor, headed by the indubitable magnetic nature of the macroscopic field of the charged little balls used by Coulomb in his famous experiments at a time when the quanta of electricity were still considered as freestanding entities, and not something attached to elementary particles with mass and own dipole magnetic field.

In addition, the spinning ring electron can be the the starting point for a new, reliable theory of the atoms formed through interactions between the dipole magnetic fields of their constituents, nuclei and electrons. But this remains a desideratum of the future, the more so as all atomic theories made until now have been based on the untimely belief in the dissociated state of molecular hydrogen passed through discharge tubes, a conviction formed in the 19th century, but in a clear and irreconcilable conflict with experimental data arising in the modern age.

www.ingramcontent.com/pod-product-compliance
Lightning Source LLC
Chambersburg PA
CBHW071431180526
45170CB00001B/296